Privacy, Security And Forensics in The Internet of Things (IoT)

Reza Montasari • Fiona Carroll • Ian Mitchell
Sukhvinder Hara • Rachel Bolton-King
Editors

Privacy, Security And Forensics in The Internet of Things (IoT)

Springer

Editors
Reza Montasari
Hillary Rodham Clinton School of Law
Swansea University
Swansea, UK

Ian Mitchell
Computer Science
Middlesex University
London, UK

Rachel Bolton-King
Law, Policing & Forensics
Staffordshire University
Stoke on Trent, UK

Fiona Carroll
Cardiff School of Technologies, Cardiff
Metropolitan University, Cardiff, UK

Sukhvinder Hara
London, UK

ISBN 978-3-030-91220-8   ISBN 978-3-030-91218-5   (eBook)
https://doi.org/10.1007/978-3-030-91218-5

This Springer imprint is published by the registered company Springer Nature Switzerland AG
The registered company address is: Gewerbestrasse 11, 6330 Cham, Switzerland

Introduction: Critical Analysis of the Challenges Police and Law Enforcement Face in Policing Cyberspace

The digital world has transformed society enabling new ways of communication and exchanging information. This innovation simultaneously poses a plethora of new challenges as cyberspace is vulnerable to extensive misuse. Technological shifts in the criminal landscape poses a myriad of challenges for policing and law enforcement that undermine the efficacy of crime control online. In view of this, this chapter provides a comprehensive account of the challenges faced by police and law enforcement in keeping cyberspace secure. This chapter will proceed in two stages. Firstly, four key challenges of policing in the digital realm will be identified: legislation, jurisdiction, anonymity and reporting. Following analysis of the challenges, this chapter will recommend possible avenues for future research to assist both in addressing the threat of cyberspace as well as the digital investigation of cybercrime. The chapter concludes that international cooperation and multi-agency partnership between state actors, private companies, academics, architects and users will provide the most advantageous response in the fight against cybercrime.

Introduction

In the twenty-first century, the internet and associate technologies such as the Internet of Things (IoT) solutions, Cloud-Based Services (CBSs), Cyber-Physical Systems (CPSs) and mobile devices have become the defining feature of modern life (Montasari & Hill, 2019; Montasari, 2017). We live in a hyper connected world that has expanded accessibility, capability and reach. The digital age has constructed a ubiquitous environment for individuals to interact, connect and share information. Technology is fundamentally advantageous for society to fuel our ability to interact across the world. However, the internet remains a double-edged sword. As the use of technology continues to grow exponentially, so do the opportunities for criminals to exploit vulnerabilities in cyberspace.

In 2016, the Crime Survey for England and Wales introduced new victimisation questions concerning fraud and computer misuse. This data revealed the unprecedented scale and nature of the problem society faces as cyber-related crime accounted for almost half of all crimes committed (Caneppele & Aebi, 2019). The threat has evolved significantly, becoming increasingly sophisticated and multifaceted, targeting not only individuals but critical infrastructure, industries and governments. Cybercrime has become one of the fastest growing types of crime in the United Kingdom (UK) and is now recognised on the National Risk Register as a Tier 1 threat to national security (Stoddart, 2016). For the purpose of this chapter, the broad umbrella term of cybercrime will be employed to include both cyber-dependant and cyber-enabled crime. The advent of technology provides cybercriminals an opportunity to change their modus operandi, this poses significant implications for policing. Traditional models of policing and law enforcement have derived from assumptions that crime occurs in physical proximity, in limited scale with traceable evidence (Harkin et al., 2018). The digital era has disrupted conventional law and understanding of enforcement as cybercrime does not operate under the same spatial and temporal constraints. Therefore, traditional policing strategies such as localised foot patrols and hierarchically organised models are not applicable to cyberspace. Whilst adversaries proved adaptive in leveraging new technological innovation to overwhelm the capabilities of state security, law enforcement agencies were ill-prepared for this transformative shift from offline to online operations. Thus, it is apparent that the domain of cybercrime is rapidly "increasing in frequency, scale, sophistication and severity" (Harkin et al., 2018).

Drawing upon an increasing body of literature, this chapter will critically evaluate the challenges police and law enforcement endure in policing cyberspace. This chapter will explore four key policing obstacles: an archaic and time-consuming legislative process, a lack of international consensus, the anonymous nature of cyberspace, and the under-reporting of cybercrime. Following critical analysis, this chapter will provide recommendations as to how to strengthen police and law enforcement responses against cybercrime. Fundamentally, the internet is a digital environment that is changing the way criminals and law enforcement operate. Given the complexities of cybercrime, there is no one solution. Therefore, this chapter will propose that no singular agency or government can police cyberspace by itself. Instead, government, industry, engineers, users, policymakers and academics must combine efforts to tackle cybersecurity challenges.

Challenges

Addressing the challenge that cyberspace represents is fraught with difficulties as new technologies and events present a myriad of legal, policy and technical work that requires months or even years to establish. Yet the pace of innovation appears relentless and cascading, threatening to overwhelm and in many cases overtake policymakers and regulatory bodies at a national and international level. Criminals by sheer innovation are redefining cyberspace which is in turn shaping and driving current approaches to cybersecurity. The following sections will outline the key challenges police and law enforcement face in establishing a sustainable and comprehensive response to criminal activity in cyberspace.

Legislation

A key challenge in policing cybercrime in contemporary society is the apparent disconnect between legislation and technology. The penal system is an inherently retroactive and lengthy process which generates numerous obstacles in regulating the cyber domain. Technological forces are evolving at a rapid rate that is far outpacing the development of policy and legislation. Consequently, cyberspace is typically governed by a patchwork of weak, under-developed and competing legislation as illustrated by the Computer Misuse Act 1990, which is the primary piece of legislation for prosecuting cyber-related offenders (Criminal Law Reform Now Network, 2020; Montasari et al., 2016). This Act was established three decades ago with no foresight into the rapid pace of technological advancements. Therefore, the concepts embodied in the Computer Misuse Act 1990 were intended to be technologically neutral, in order to pertain to both current and future technologies. Despite attempts to future-proof, the emergence of unprecedented technology has created a number of loopholes and ambiguity in the application of the law. Consequently, Ashworth (2013) argues that legislation merely sustains a myth of control and legitimacy of the sovereign state, as legislation is often impromptu, expressive and inconsistent in an attempt to cater to public pressures with little consideration of an evidence-base and expert knowledge.

Private and public sectors are typically eager to spend money, time and resources into the enforcement of computer misuse laws in terms of the apprehension and prosecution of offenders. However, the legal foundation is sorely lacking in substance. Legislation has demonstrated an inability to target the right people and establish defences which enable private enforcement to work effectively alongside public enforcement in order to best address cyber threats. Thus, the cybersecurity industry remains constrained and inadvertently criminalised by the Computer Misuse Act (Criminal Law Reform Now Network, 2020). For this reason, critics, namely the Criminal Law Reform Now Network (2020), have claimed that the Computer Misuse Act 1990 is outdated and does not reflect the current problems police and law enforcement face. Subsequently, they deduce that cyber legislation needs to undertake radical reform. Evidently, establishing legislation to coincide with ever-evolving technology proves challenging. Despite uncertainties, it is critical to not wholly dismiss the role of law and policy on the basis that technology will always evolve more rapidly (Sallavaci, 2017). Ultimately, legislation remains a crucial component in the fight against cybercrime. An Act of Parliament provides police and law enforcement with a fundamental blueprint that guides behaviour and establishes standards and frameworks. Nonetheless, in an ever-changing highly digitised realm, substantial amendments to the rule of law are necessary to evolve in line with society.

Jurisdiction

The infrastructure of the internet is a physical construct that exists in time and space within physical borders of sovereign countries. However, the data flowing throughout this infrastructure spans across multiple national jurisdictions, which remains an inherent challenge of cyberspace. Whilst criminal activity in cyberspace penetrates effortlessly across geographical borders, law enforcement does not. As a result, nationally bounded law enforcement is required to operate within a realm that is geographically unbounded, thereby evoking a large number of complications (Kennedy & Warren, 2020). The most prominent international instrument concerning cybercrime is 'The Council of Europe Convention on Cybercrime', also known as the Budapest convention. The convention seeks to harmonise national laws on cybercrime, improve investigative techniques and increase cooperation between nations (Kennedy & Warren, 2020). However, achieving consensus proves a contentious issue as each nation possesses their own independent norms, beliefs and practices, and thus promote differing visions for cyberspace. For instance, various governments advocate for cyber sovereignty contending that national borders apply to cyberspace and each country should have the right to govern how people and businesses use the internet within their territory. Whereas other nations support internet freedom, the concept that every citizen should be free to express themselves and spread new ideas online with anyone, anywhere (Kennedy & Warren, 2020).

This fragmentation between nations renders it almost impossible to establish an international consensus concerning internet governance and regulation. However, this is not to say the Budapest Convention as a whole is obsolete. Despite its limitations, the treaty provides a fundamental framework in facilitating international cooperation and the harmonisation of legislation. Evidently, what makes cybercrime difficult to monitor and enforce is its transgressive form, one that does not respect international borders. The internet and computers have enabled individuals to steal electronic data remotely without physical proximity. Thus, criminal actors operating across borders adds a level of complexity to policing as victims, perpetrators and evidence can all reside within differing jurisdictions (Montasari, 2017). Consequently, police forces must request data preservation and access to electronic evidence residing in other jurisdictions. This reliance upon mutual assistance makes it incredibly complex, time consuming and costly to bring offenders to justice outside of the United Kingdom. As a result, recent studies, including Świątkowska's, call (2020) for more effective and synchronised international efforts to mitigate digital vulnerabilities. She determines that a lack of international consensus can offer cybercriminals a spatial safe haven whereby they operate outside the scope of law enforcement and international legislation. These safe havens provide a domain for adversaries to better evade government restrictions, detection and prosecution. Amid the global disagreement, technological innovation continues to accelerate at a tremendous speed. Therefore, international cooperation is vital to eradicate the safe haven for cyber criminals, promote information sharing and eventually enhance global investigations.

Anonymity

A further challenge the police face in the apprehension of cyber criminals is anonymity. There are many publicly available and accessible tools that allow users' internet activity to remain anonymous. The most commonly used anonymous system is the Tor browser, this is a powerful tool that offers online end-to-end encryption through masking a user's IP address (Davies, 2020). This offers the ability to protect privacy and effectively prevent governments from accessing data and tracking online activities. This freedom from censorship is therefore deemed by civil rights activists as a powerful tool to be utilised in heavily monitored and authoritarian states. Whilst encrypted communication protects the security and privacy of its users, it also presents significant disadvantages as users of illegal sites leverage this cloak of anonymity to evade police and law enforcement. There remain a series of websites hidden under a layer of protection that can only be accessed utilising specialised anonymous browsers. This realm has been deemed the dark net. Criminals can mask their identities and hide their locations by redirecting communication and activity through a distributed network of relays around the world. Whilst the dark web is not exclusively used by criminals, these hidden services can create a centralised repository of illicit marketplaces facilitating the selling and distributing of illegal goods such as firearms, drugs, counterfeit currency and child pornography (Davies, 2020). Consequently, the nature of cyberspace is problematic for policing as the risk of apprehension can be easily mitigated through utilising Tor browsers, cryptocurrency and virtual private networks.

The dark net is constantly evolving and adapting as these illicit markets operate on the fringes of the internet and are quick to adopt readily available technology in order to provide greater anonymity. This is exemplified in Ladegaard's research (2019) into the most prolific dark net investigation, Operation Onymous. Ladegaard (2019) reported that criminals will typically migrate to alternate cryptomarkets once their current darknet market is detected and removed by law enforcement. From these findings there is evidence to suggest that the cybercrime ecosystem is resilient to law enforcement takedowns as operations merely lead to a displacement of criminal activity. Arguably police crackdowns can trigger criminal innovation as infiltration forces darknet markets to enhance their security and infrastructure. Overall, anonymity in cyberspace remains a significant challenge for police investigations as criminals continue to circumvent government surveillance and detection.

Reporting

There is a vast amount of crime that goes unnoticed, unreported and undetected. This generates what scholars term, the dark figure of crime (Kemp et al., 2020). Action Fraud is the centralised reporting agency of fraud and cyber offences. However, according to the Office for National Statistics (2020) only 338,255 cases of fraud and cyber-crime were recorded by Action Fraud within a 12-month period. Whereas the Crime Survey for England and Wales recorded approximately 4.5 million incidents (Office for National Statistics, 2020). This reveals that only 7% of victims reported incidents of cybercrime and fraud to the police, as such there remains a large discrepancy between what people experience and what they report to the police. Therefore, police recorded crime does not represent the true nature and scale of the cyber problem the United Kingdom is facing. This variation between statistics highlights the advantages of victim surveys to shed light upon the dark figure of crime and the severe limitations associated with relying upon police-recorded data (Kemp et al., 2020).

There remains a significant problem with under-reporting within the realm of cybercrime as it depends upon a victim's willingness to report a crime. There are a multitude of reasons why individuals and businesses may not report a cybercrime to the police for instance, a lack of awareness of victimisation, fear of stigma, poor reporting mechanisms and potential reputation damage (Bailey et al., 2021). Cybercrime does not always have a readily identifiable victim, and it may be difficult to determine and recognise one's own victimisation, consequently computer-related crime is often referred to as 'hidden crime'. Moreover, even supposing an individual is aware of their own victimisation, they may feel too embarrassed or ashamed to report the incident. This notion is evidenced by several academics, including scholars Bailey et al. (2021), who determined that victim blaming discourse permeates cybercrime. Findings from in-depth qualitative interviews found that cyber victims frequently view themselves as partly to blame for their victimisation as participants often referred to themselves as 'gullible', 'stupid' and 'naïve'. Many participants suffered from severe psychological harm including anxiety and para-noia and experienced a breakdown of personal relationships following victimisation. Subsequently, internalised and externalised stigma may seek to explain the high levels of underreporting within cyber-related crime. It is important to note that (Bailey et al., 2021) dataset pertains to a small sample size of 80 victims; despite this methodological limitation, the study provides a rich insight into the lived experiences of cyber victimisation and the challenges of reporting cybercrime.

Official statistics that represent an accurate figure of crime are an important aspect of police operations as data can assist in detecting trends and patterns amongst criminal activity. Therefore, data analysis can help inform financial budgets and resource allocation to ensure police interventions are implemented successfully and effectively. Police ought to coordinate activity and focus their enforcement resources upon problem areas; however, with a limited dataset due to under-reporting, this proves challenging (Caneppele & Aebi, 2019).

Future Direction

Given the compounding challenges police face in cyberspace, a nodal network of regulation is required that combines private and public, state and nonstate, national and international institutions. A multifaceted threat requires a multi-layered, global, dynamic and decentralised regulatory system in order for the problem to be addressed. The following sections will offer recommendations to improve cyber protection, investigations and response.

Legistlation Reform

The Criminal Law Reform Now Network (2020) determines that legislation is "crying out for reform". In the Computer Misuse Act 1990, judicial lexicon remains broad and notoriously vague; subsequently, it permits a vast amount of flexibility in the application of the law. However, exercising prosecutorial discretion may result in inconsistent and unjust rulings. As evidenced in the case of R v Cuthbert, a computer security consultant was convicted for performing unauthorised penetration testing on a suspected inauthentic website. This ruling sparked many concerns in the penetration testing community due to fears that the law makes no distinction between good faith and malicious intent (Criminal Law Reform Now Network, 2020). Consequently, Guinchard (2021) echoes the Criminal Law Reform Now Network (2020) and proposes a radical reform of cyber legislation. Currently, an individual can be prosecuted under the Computer Misuse Act without the requirement for malicious intent; therefore, the act invertedly criminalises cyber security researchers. Most notably, the making, supplying or obtaining of hacking tools equates a computer misuse offense which inhibits vulnerability testing and threat research. Therefore, Guinchard's (2021) chapter recommends the introduction of a 'public interest' defence to allow detected vulnerabilities in systems and networks to be safely disclosed without fear of legal persecution. Guinchard's argument is persuasive as reform of the Computer Misuse Act 1990 to include a public interest defence can enable more freedom for security professionals to investigate vulnerabilities in critical national infrastructure.

The Covid-19 pandemic has highlighted the increasing need for a modernised legislative framework for law enforcement as society becomes ever more reliant upon digital technology. The pandemic saw a rapid acceleration and significant uptake of individuals around the world working from home. This greatly increased the potential pool of victims as a number of companies and individuals struggled to provide rapid security and infrastructure. Criminal organisations attempted to capitalise upon this unforeseen shift as new vulnerabilities surfaced from remote working (Buil-Gil et al., 2021). The pandemic demonstrates the need to establish adaptable and resilient judicial responses as the nature of the cyber threat is dynamic and evolving at an alarming rate.

Multi-Agency Response

Cybercrime is inherently networked and sophisticated thus the nature of the threat demands an integrated and collective regulatory response. Therefore, the policing of cyberspace calls for a multi-agency layered approach to establish a comprehensive and decentralised defence framework. Therefore, internet governance must operate seamlessly between public and private sectors, state and non-state actors, and national and international bodies. However, coordinating a sustainable and efficient collaborative effort when different organisations and administrations have differing agendas is a complex process (Leppänen & Kankaanranta, 2020). Typically, national security strategies have overlooked the role of private industry as essential stakeholders. However, private entities own and operate the infrastructure within cyberspace and are often the victim of cybercrime; therefore, it is imperative to incorporate the private sector into the policing of cyberspace. In the United Kingdom, the National Cyber Security Centre (NCSC) is a central hub of expertise that provides a significant foundation in improving public-private collaboration; however, much more work remains to be done (Stoddart, 2016). The NCSC was established to simplify the landscape for cybersecurity and devise a single point of contact. In doing so, the NCSC harmonises the way law enforcement and the private sector communicate with one another to better detect threat actors and conduct better investigations (Stoddart, 2016). Whilst the work the NCSC do has enhanced collaboration and information sharing, it is necessary to build upon this further by encouraging greater cooperation between organisations and law enforcement to better disrupt cyber criminals. This can be accomplished through a modernised legal and regulatory framework that encourages multi-agency collaboration.

The United Kingdom lives in an era whereby digital evidence is rampant in nearly every crime. Despite this, there remains tension between the transnational horizontal nature of the internet and the vertical structure of the United Kingdom's jurisdictional system based upon the geographical conception of nation states with distinct borders. Cybercrime is truly a global problem; it has no respect for traditional police force boundaries. This new era of connectivity underscores the need for international arrangements that encourage responsible cyber practices. A fundamental aspect that requires development is the capacity to exchange information amongst private and public entities across jurisdictions. Cybercriminals have the ability to operate in a flexible and agile way across borders; however, law enforcement remain restricted to local jurisdictions (Leppänen & Kankaanranta, 2020). As has already been highlighted, cybercrime is not a closed border issue, policymakers and academia must view this domain from an international perspective.

Evidence-Based Policing and Training

Police and law enforcement who attempt to solve issues of cyberspace in the sense of policymaking, legislation and enforcement often lack knowledge of the space they are regulating. Therefore, in a domain that is technologically diverse and dynamic, collaboration between police agencies and the academic sector is critical to accumulate a comprehensive evidence base. Evidence-based approaches bring a powerful tool of systematic analysis, evaluation, testing and empirical studies to policing. Evidence-based policing focuses upon knowledge that is derived from rigorous evaluations of new and existing tactics and strategies (Koziarski & Lee, 2020). Thus, this is a concept that comes from partnership between researchers and practitioners to understand the relationship between action and outcomes. Evidence-based policing has increasingly permeated UK police forces; despite this, responses to cybercrime remain an underdeveloped domain. Knowledge of 'what works' or does not work in policing cyberspace is scarce (Koziarski & Lee, 2020). As policing becomes more complex, mechanisms of oversight and scrutiny will become increasingly important to guarantee a significant degree of public trust and confidence. Therefore, it is essential for governing bodies in collaboration with researchers to evaluate and review cybercrime policing approaches in order to determine the most effective strategies for law enforcement to implement.

Cyber criminals have developed an integrated and sophisticated web of skills; therefore, investigations are complex and require specialist tools and skillsets. As a result, cybercrime requires augmenting the skill set within the police and judicial system at all levels to meet this changing environment. However, typically, law enforcement has not been well equipped to deal with emerging threats and the increasing demands placed upon it. The majority of literature concerning cyber-crime and policing acknowledge training as a prevalent issue for staff; however, there remains a shortage of detailed insight. In order to expand the knowledge base, Schreuders et al. (2018) conducted in-depth interviews with officers from a United Kingdom police force. They subsequently found that officers did not possess the necessary skills or technological background required for everyday digital investigations. This chapter indicates that there must be an upskilling of police officers beyond cyber-specific agencies as cybercrime is a wider problem that intersects all types of crime. Thus, cyber knowledge and awareness needs to transpire across the core of police activity.

This claim is supported by the National Police Chief's Council (2016) in their Policing Vision 2025 report whereby they acknowledge that the advances in digital technology are presenting significant challenges and opportunities to policing. The chapter calls for transformative change and outlines a vision as to how the incorporation of technology in policing can address current and future threats in the digital era (National Police Chiefs Council, 2016). As society digitally evolves it becomes increasingly important for law enforcement agencies to be equipped with the appropriate skills, knowledge and investigative capabilities to leverage this technology. Police forces must enhance their ability to train and upskill existing personnel to meet this changing environment and capitalise upon an existing knowledge base within the workforce. Ultimately, this chapter recognises the requirement for evidence-based digital policing to permeate wider police training and tactical strategies moving forward.

Public Awareness

There is a growing role of human factors in shaping the likelihood of cyber security breaches as users are the gatekeepers of sensitive data and systems. Understanding how human error shapes the threat landscape is therefore vitally important in attempting to mitigate cybercrime (Monteith et al., 2016). In line with this notion, Williams (2016) determines that cyber defence rests upon the commitment of every citizen and thus recommends a radical overhaul of conventional reactive policing methods. Williams analysed Eurobarometer survey data and suggests that routine activity theory is applicable to the conditions of cyberspace as users' online conduct can influence the commission of an offence. Routine activity theory determines that a crime is likely to occur with the convergence of a suitable target, potential offender and absence of a capable guardian. Consequently, within cyberspace, individuals can mitigate their risk of victimisation by employing passive guardianship measures in the form of secure browsers and antivirus software (Williams, 2016). Ensuring the public are educated in how to protect their devices appropriately can increase cybersecurity, thereby decreasing the need for police intervention. However, a key challenge lies in ensuring that citizens understand the significance of cyber threats and the role individual users play in cyber security.

In order to encourage user compliance, Brenner (2007) proposes a new punitive crime-control strategy that relies upon self-policing and user 'responsibilisation'. She determines that individual users should be held liable for their own cyber-security under criminal law. Therefore, if a victim fails to implement up-to-date security measures in order to protect one's own computer system, they will no longer be entitled to a response from law enforcement. Brenner (2007) extends this principle and determines that users who harm others as a result of their own lack of security measures should be found liable of a criminal offence under the principle of negligence. Whilst it is critical to encourage users to prevent their own victimisation, this punitive approach remains fundamentally flawed as it is rooted in notions of victim blaming. Denying cyber victims the right to a police investigation overlooks offenders' culpability and places the onus entirely upon victims. As aforementioned, there is now a considerable body of research which suggests that victims often encounter shame and stigma when reporting cybercrime; therefore, Brenner's framework (2007) perpetuates the notion that victims are to blame for not adequately protecting themselves. Instead, victims of cybercrime should be supported and offered resources to prevent any future victimisation (Monteith et al., 2016).

For future research and interventions, it is imperative to diversify cyber security beyond traditional law enforcement to consider the implications of user behaviour and action. A poor understanding of technology and its vulnerabilities puts users and companies at risk. Therefore, human factors present an opportunity for making systems safer, more robust and more resilient. Thus, there is reason to conclude that designing public awareness campaigns to educate communities on the dangers of cyberspace and develop cyber skills can help build resilience to crime in an increasingly digital world. Ultimately, a comprehensive strategy must focus not only upon the apprehension of offenders and legal pursuit but also the prevention of victimisation.

Conclusions and Recommendations

Evidently, the ecosystem of internet governance is a multifaceted issue with no singular solution. Due to the sheer volume and breadth of cyberspace, police and law enforcement efforts alone cannot fully address the challenge of cybercrime. Cybercrime is not an area that can be comprehensively tackled by an exclusive focus on cybercrime as a legal, policy or technical problem, but rather an understanding of these individual domains requires an understanding of the others. With the proliferation of new technology formulating new capabilities throughout homes, namely the advancement of artificial intelligence and the Internet of Things, it stands with good reason that cybercrime will continue to escalate in the near future. This illuminates the imminent need for policing and law enforcement practices to evolve in line with technological advances to enhance cyber-resilience within critical infrastructures. As the threat landscape is growing in complexity, there are fundamental hurdles in addressing cybercrime. This chapter has investigated the core challenges for policing cyberspace; cybercrime transgresses jurisdictional boundaries, provides anonymity, creates legislative ambiguity and experiences high levels of under-reporting. Consequently, police and law enforcement require an innovation revolution that enables the workforce to evolve with an increasingly digitised and networked society. Moving forward, preventative measures to increase community resilience and user responsibility ought to be accompanied by a skilled criminal justice taskforce to investigate and prosecute offenders at a regional and international level. Alongside this, there is a need for evidence-based cyber policing to inform and evaluate strategies and ensure practices are fundamentally rooted in an effective knowledge base. This chapter has critically evaluated the avenues for future research and work in the policing of cyberspace. Ultimately, as society moves forward, there is a need for national and international collaboration between private companies, public agencies, academia and users to ensure a robust and effective response to cybercrime.

Hillary Rodham Clinton School of Law,   Aime Sullivan
Swansea University, Swansea, UK

Hillary Rodham Clinton School of Law,   Reza Montasari
Swansea University, Swansea, UK

References

Ashworth, A. (2013). *Positive obligations in criminal law*. Bloomsbury Publishing.

Bailey, J., Taylor, L., Kingston, P., & Watts, G. (2021). Older adults and "scams": Evidence from the mass observation archive. *The Journal of Adult Protection*. http://hdl.handle.net/10034/624222

Brenner, S. W. (2007). Cybercrime: Re-thinking crime control strategies. In Y. Jewkes (Ed.), *Crime online* (pp. 12–28). See ncj-218881.

Buil-Gil, D., Miró-Llinares, F., Moneva, A., Kemp, S., & Díaz-Castaño, N. (2021). Cybercrime and shifts in opportunities during covid-19: A preliminary analysis in the UK. *European Societies, 23*(Suppl. 1), S47–S59.

Caneppele, S., & Aebi, M. F. (2019). Crime drop or police recording flop? On the relationship between the decrease of offline crime and the increase of online and hybrid crimes. *Policing: A Journal of Policy and Practice, 13*(1), 66–79.

Criminal Law Reform Now Network. (2020). Reforming the Computer Misuse Act 1990. http://www.clrnn.co.uk/media/1018/clrnn-cma-report.pdf. Accessed June 07, 2021.

Davies, G. (2020). Shining a light on policing of the dark web: An analysis of UK investigatory powers. *The Journal of Criminal Law, 84*(5), 407–426.

Guinchard, A. (2021). The criminalisation of tools under the computer misuse act 1990. The need to rethink cybercrime offences to effectively protect legitimate activities and deter cybercriminals. In *Rethinking cybercrime* (pp. 41–61). Springer.

Harkin, D., Whelan, C., & Chang, L. (2018). The challenges facing specialist police cyber-crime units: An empirical analysis. *Police Practice and Research, 19*(6), 519–536.

Kemp, S., Miró-Llinares, F., & Moneva, A. (2020). The dark figure and the cyber fraud rise in Europe: Evidence from Spain. *European Journal on Criminal Policy and Research, 26*(3), 293–312.

Kennedy, S., & Warren, I. (2020). The legal geographies of extradition and sovereign power. *Internet Policy Review, 9*(3), 1–18.

Koziarski, J., & Lee, J. R. (2020). Connecting evidence-based policing and cybercrime. *Policing: An International Journal, 43*(1), 198–211.

Ladegaard, I. (2019). Crime displacement in digital drug markets. *International Journal of Drug Policy, 63*, 113–121.

Leppänen, A., & Kankaanranta, T. (2020). Co-production of cybersecurity: A case of reported data system break-ins. *Police Practice and Research, 21*(1), 78–94.

Montasari, R. (2017). An overview of cloud forensics strategy: Capabilities, challenges, and opportunities. In *Strategic Engineering for Cloud Computing and Big Data Analytics* (pp. 189–205).

Montasari, R., & Hill, R. (2019). Next-generation digital forensics: challenges and future paradigms. In *2019 IEEE 12th International Conference on Global Security, Safety and Sustainability (ICGS3)* (205–212). IEEE.

Montasari, R., Peltola, P., & Carpenter, V. (2016). Gauging the effectiveness of computer misuse act in dealing with cybercrimes. In *2016 International Conference on Cyber Security and Protection of Digital Services (Cyber Security)* (pp. 1–5). IEEE.

Monteith, S., Bauer, M., Alda, M., Geddes, J., Whybrow, P. C., & Glenn, T. (2016). Increasing cybercrime since the pandemic: Concerns for psychiatry. *Current Psychiatry Reports, 23*(4), 1–9.

National Police Chiefs Council. (2016). Policing Vision 2025. https://www.npcc.police.uk/documents/Policing%20Vision.pdf. Accessed June 07, 2021.

Office for National Statistics. (2020). Crime in England and Wales: Year ending March 2020. https://www.ons.gov.uk/peoplepopulationandcommunity/crimeandjustice/bulletins/crimeinenglandandwales/yearendingmarch2020. Accessed June 07, 2021.

Sallavaci, O. (2017). Combating cyber dependent crimes: The legal framework in the UK. In *International Conference on Global Security, Safety, and Sustainability* (pp. 53–66). Springer.

Schreuders, Z. C., Cockcroft, T. W., Butterfield, E. M., Elliott, J. R., Soobhany, A. R., et al. (2018). Needs assessment of cybercrime and digital evidence in a UK police force. *International Journal of Cyber Criminology, 14*(1), 316–340.

Stoddart, K. (2016). Uk cyber security and critical national infrastructure protection. *International Affairs, 92*(5), 1079–1105.

Świątkowska, J. (2020). Tackling cybercrime to unleash developing countries' digital potential. In *Pathways for Prosperity Commission on Technology and Inclusive Development* (p. 2020–01).

Williams, M. L. (2016). Guardians upon high: An application of routine activities theory to online identity theft in Europe at the country and individual level. *British Journal of Criminology, 56*(1), 21–48.

Contents

Part I
Privacy, Security and Challenges in the IoT

Ethics and the Internet of Everything: A Glimpse into People's Perceptions of IoT Privacy and Security

Fiona Carroll, Ana Calderon, and Mohamed Mostafa

1 Introduction

The Internet of Things (IoT) can be described as an agglomeration of 'things' that are embedded with sensors and other technologies in order to connect and share data with other devices across the Internet. Nowadays, with the availability of cheap sensors, IoT enables various devices and objects around us to be addressable, recognizable and locatable (Atlam & Wills, 2020). And it is this networked scenario that is hugely impacting our society, work and life. For example, IoT has opened up a range of new opportunities and experiences for us, and it has made us more efficient in work and has made us safer in our homes and vehicles. However, as van Deursen et al. (2019) describe the daily use of IoT does not require extensive user skills (i.e. IoT operates 'on its own') and once these devices become part of an interconnected system in which they are connected to a multitude of other devices, the story gets more complex. Indeed, IoT is changing the ways people, businesses and governments interact among themselves (Economides, 2017). And as the authors of this chapter have found, it is not always a change for the greater good of society and humanity.

This chapter will take a look at users' perceptions around IoT whilst exposing the need for a trust framework to enforce ethical behaviours (i.e. ownership, trust and accountability), privacy and security and appropriate use of IoT in networked environments. The first section reviews the ethics of IoT. Following that, the chapter documents two studies: study 1 conducted a survey investigating the perceptions of personal data in the digital age which allowed for statistical as well as qualitative analyses, and study 2 utilized social networks to extract people's views of IoT and

F. Carroll (✉) · A. Calderon · M. Mostafa
Cardiff School of Technologies, Cardiff Metropolitan University, Cardiff, UK
e-mail: fcarroll@Cardiffmet.ac.uk; acalderon@Cardiffmet.ac.uk; mmostafa@Cardiffmet.ac.uk
https://www.cardiffmet.ac.uk/technologies/Pages/default.aspx

R. Montasari et al. (eds.), *Privacy, Security And Forensics in The Internet of Things (IoT)*, https://doi.org/10.1007/978-3-030-91218-5_1

privacy. The chapter concludes with a discussion on the main points of interest from the studies and then an overview of the bigger IoT picture. In particular, how IoT is not only transforming the sphere of big businesses of today but also the impact (positive and negative) it is having in people's daily lives.

2   The Ethics of IoT

We cannot deny that IoT offers great benefits to productivity; however, as Williams et al. (2018) highlights, IoT is also increasingly pervading our lives. We are seeing more and more of our critical societal services (CSSs) that provide electricity, water, heat and ways to travel, communicate and trade (i.e. vital systems) becoming part of the Internet of Things (IoT) (Asplund & Nadjm-Tehrani, 2016). And in this IoT scenario, the satisfaction of security and privacy requirements, such as data confidentiality and authentication, access control within the IoT network and privacy and trust among users and things, and the enforcement of security and privacy policies need to play a fundamental role (Sicari et al., 2015). Interestingly, in their paper, Zheng et al. (2018) highlight several recurring themes, one of which centres around users' desires for convenience and connectedness and how these desires dictate their privacy-related behaviours for dealing with external entities, such as device manufacturers, Internet Service Providers, governments and advertisers. Essentially, as IoT is built on the basis of the Internet, security problems of the Internet will also show up in IoT (Tewari & Gupta, 2020).

A core aspect of this lies in the fact that IoT collects and deals with unprecedented volumes of private, real-time and detailed data (AlHogail, 2018). But what happens with this data, what happens to our privacy and security around this data? In the midst of all this unprecedented amount of data being collected, Mashhadi et al. (2014) raise an important question: who owns this data and who should have access to it? From an end users perspective, it is hard to see and understand the scale of the full IoT picture. As van Deursen et al. (2019) describe ownership can be ascribed to a relatively limited set of devices: activity trackers, heart rate monitors, sport watches, smart thermostats and lightning systems. However, in reality, how many other million devices are collecting information on us? There is no doubt that trust management needs to play an important role in IoT for reliable data fusion and mining, qualified services with context awareness and enhanced user privacy and information security (Yan et al., 2014). However, in their research, Alraja et al. (2019) showed the trust in the IoT was also affected by both the users' risk perception and their attitudes towards using the IoT.

Thus, it creates, as Tzafestas (2018, p. 1) describes 'a new social, economic, political, and ethical landscape that needs new enhanced legal and ethical measures for privacy protection, data security, ownership protection, trust improvement, and the development of proper standards'. Indeed, the world of IoT has huge potential to enhance society, but it has all the traits that could also destroy it.

2.1 Ownership

What does ownership really mean in our online world? For example, when we buy a movie and/or music from iTunes, Amazon or other, do we actually own this item? As Rosenblatt (2016, p. 1) highlights 'the answer is mostly no: you are licensing it on some terms that the retailer sets, which usually don't amount to ownership'. To further complicate matters, IoT is enabling new digitally enhanced products and has the power to alter existing products in such different ways. Indeed, buying digital media is not the same as buying a physical book, what you are buying is a lifetime licence to use this digital file rather than a physical tangible asset. Unknown to many of us, the digital book or film or song is not our property to sell or share with others. In that sense, the experience of buying an asset in the physical is very different to that of the virtual world. Moreover, the concept of ownership and what constitutes property does not fully square up with what we know and experience in the physical world. In detail, ownership can be described as the state of having complete legal control of the status of something (i.e. with the right to transfer possession to others), whilst property is something that is owned.

In the virtual IoT world, when we buy a digital book, we do not have the same right of ownership and it is a different type of property when compared to the buying of that book in the physical shop. As Stallings (2011) asserts, there are three primary types of property. The first is what he calls real property which is something we are all very familiar with in the physical world; this includes land and things permanently attached to the land such as trees, buildings etc. The second is personal property which can be personal effects, moveable property and goods such as cars, bank accounts, wages securities etc. And, the third is intellectual property which is any intangible asset that consists of human knowledge and ideas such as software, data, novels, sound recordings etc. So, in terms of the physical and digital book, we see a shift from the personal property to the intellectual property and with this comes different ownership rights. In their book *The End of Ownership: Personal Property in the Digital Economy*, Perzanowski and Schultz (2016, p. 1) highlight that this is a problem, they question what property really means in the online environment and what should be done about it in law. Interestingly, in 2006, it was described as a bigger problem where the solutions need to come not just from the law itself but also from three other modalities of regulation: technology, the market and behavioural norms (Lessig, 2006).

Regardless, what is of interest to the authors of this chapter is that 'most consumers are poorly informed about the disparities between ownership and licensing', vary—from retailer to retailer, from publisher to publisher and from product to product (Perzanowski & Schultz, 2016, pp. 6–7). In terms of IoT, this is complicated even further as there is a great number of players involved in the process of generating, collecting and processing IoT. It becomes a complicated legal question (i.e. who will be the owner of such data and hence who is legally entitled to carry out 'business' with such data etc.). In line with this, we ask the question of do we even have ownership of our own privacy with IoT? In the physical world, retailers generally cannot easily keep track of who buys, owns, shares or resells a physical book. However, the nature of IoT as well as 'the architecture of online media allows unprecedented surveillance of consumer behaviour' (Perzanowski & Schultz, 2016, p. 7).

2.2 Trust

As the Internet of Things (IoT) matures, it continues to have a profound effect on businesses and business models. However, it is becoming clear that IoT's success for businesses really depends on the level of trust consumers have in it. Consumer confidence is crucial for IoT to thrive. Yet, many of today's digital products and services are rushed to market at the lowest possible cost with little consideration for people's basic security and privacy protections (Online Trust Alliance, 2017, p. 1). Beyond a doubt, consumers today need to trust businesses not only to collect, store and use their digital information in a manner that is of value to them but also to protect them. The Online Trust Alliance (OTA, 2017) is an initiative of the Internet Society that aims to achieve this by raising the level of security for IoT devices to better protect consumers and the privacy of their data. As Hudson (2018, p. 15) highlights, we have reached a point where we are needing TIPPSS for IoT: 'Trust (allow only designated people or services to have device or data access); Identity (validate the identity of people, services, and "things"); Privacy (ensure device, personal, and sensitive data are kept private); Protection (protect devices and users from physical, financial, and reputational harm); Safety (provide safety for devices, infrastructure, and people); and finally Security (maintain security of data, devices, people etc.)'.

It goes without saying that to fully utilize the potential capabilities of IoT, trust existence among these connected 'things' is essential and traditional security measures are not enough to provide the comprehensive security to this connected world (Altaf et al., 2019). As Voas et al. (2018, p. 1) highlight many 'IoT devices interact with the physical world in ways conventional IT devices usually do not and many IoT devices cannot be accessed, managed, or monitored in the same ways conventional IT devices can'. Furthermore, recently Mohammadi et al. (2019) found that despite a crucial demand for a trust model to guarantee security, authentication, authorization and confidentiality of connected things, in their study, they also identified several hardware and software challenges which remain unsolved due to heterogeneity essence and incremental growth in the number of IoT nodes. In further detail, Meng (2018) showed that the IoT allows smart objects to be sensed and controlled remotely under certain network frameworks, and as a result cyber criminals can hijack the communication among sensors or directly control a sensor by spreading malicious applications. Moreover, as Caminha et al. (2018) note, attacks such as the On–Off attack can threaten the IoT trust security through nodes performing good and bad behaviours randomly, to avoid being rated as a menace.

To counter these vulnerabilities, cutting edge research such as blockchain technologies are being explored to address these challenges by providing a tamper-proof audit trail of supply chain events and data associated with a product life cycle (Barakati & Almagwashi, 2020). However, it does not solve the trust problem associated with the data itself, and as Barakati and Almagwashi (2020) point out, reputation systems might be the answer (i.e. might be a more effective approach to solve this trust problem). Also, in their paper, Kotis et al. (2018) talk about supporting the selection and deployment of IoT entities based on the notion of trust semantics, using fuzzy ontologies to serve as a secure selection key to an IoT application (or service) for selecting the entities that the application should trust for its effective deployment in the specific environment/context. Despite movement to solve the trust problems, the high-profile attacks, combined with uncertainty about security best practices and their associated costs, are still keeping many businesses from fully adopting the technology. As a recent survey (IS, 2019) shows, 75% of people distrust the way data is shared and 63% of people find connected devices 'creepy'. Interestingly, a high number think that privacy and security standards should be assured by regulators (88%), followed by manufacturers (81%) and championed by retailers (80%) (IS, 2019).

2.3 Accountability

As we have seen, the complex ecosystem surrounding IoT devices means trusting IoT is not a given; the diversity and heterogeneity of components in the IoT makes it challenging to build it both secure and accountable. In terms of businesses, many of the potential benefits will not be fully realized unless people are comfortable with and embrace the technologies. As Singh et al. (2018, p. 1) highlight, 'accountability is crucial for trust, as it relates to the responsibilities, incentives, and means for recourse regarding those building, deploying, managing, and using IoT systems and services'. Yet the physical, ubiquitous and autonomous nature of IoT naturally lends itself to various accountability challenges relating to safety and security, privacy and surveillance, and governance and responsibility (Singh et al., 2018). In detail, Urquhart et al. (2019, p. 1) point out the need to build accountability into the IoT is motivated by 'the opaque nature of distributed data flows, inadequate consent mechanisms and lack of interfaces enabling end-user control over the behaviours of Internet-enabled devices'.

Interestingly, in their longitudinal study, Jakobi et al. (2018) report on a design case study in which they equipped twelve households with DIY smart home systems for two years and studied participants' strategies for maintaining system awareness, from learning about its workings to monitoring its behaviour. They found that people's needs regarding system accountability changed over time. Their privacy needs were also affected over the same period. They found that participants initially looked for in-depth awareness information, but in the later phases, their focus was on the system only when things were 'went wrong'. In terms of system accountability, they found that 'a system's self-declaration should focus on being socially meaningful rather than technically complete, for instance by relating itself to people's activities and the home routines' (Jakobi et al., 2018, p. 1).

Accountability can be described as key to building consumer trust and is mandated by the European Union's general data protection regulation (GDPR) (Crabtree et al., 2018). However, how do we enforce and encourage best practice on the ground? As Ciardiello and Di Liddo (2020) suggest, penalties provide incentives to data sharing since they redistribute firms' responsibility against data breaches. Also, the Internet Engineering Task Force (IETF) has taken the initiative to bring a standard (RFC8520), which will encourage manufacturers of IoT devices to provide a Manufacturer Usage Description (MUD) for their IoT devices (Yadav et al., 2019). Whilst there is still much work to be done to tackle the accountability IoT challenge, the authors of this chapter are confident that we are starting to move in the right direction.

2.4 Privacy and Security

Privacy and security are among the significant challenges of the Internet of Things (IoT). As Tawalbeh et al. (2020, p. 1) highlight 'improper device updates, lack of efficient and robust security protocols, user unawareness, and famous active device monitoring are among the challenges that IoT is facing'. Part of the problem is that information about the privacy and security of IoT devices is not readily available to consumers who want to consider it before making purchase decisions. Whilst legislators have recommended 'adding succinct, consumer accessible, labels, they do not provide guidance on the content of these labels' (Emami-Naeini et al., 2020, p. 1).

As the IoT ecosystem continues to grow, it has never been more urgent to prevent IoT from causing an unacceptable risk of human injury or physical damage. It is important to consider social behaviour etc. time to consider social behaviour and ethical use of IoT technologies to enable effective security and safety (Atlam & Wills, 2020). However, in their paper, Alshohoumi et al. (2019) present findings that disclose that none of the IoT architectures that they investigated considered privacy concerns which they feel needs to be considered as a critical factor of IoT sustainability and success. In fact, they stress the inevitable need to consider security and privacy solutions when designing IoT architecture (Alshohoumi et al., 2019). Furthermore, Zhou et al. (2019, p. 1) acknowledge that 'IoT has caused acute security and privacy threats in recent years', and despite there being increasing research works to ease these threats, many problems still remain open.

Looking at some of this research, Alfandi et al. (2021) investigate blockchain technology as a key pillar to overcome many of IoT security and privacy problems. Yao et al. (2020) clarify the complicated security and privacy issues and divide the life cycle of a physical object into three stages of pre-working, in-working and post-working. On this premise, they put forward a physical object-based security architecture for the IoT (Yao et al., 2020). For example, a physical object often needs to communicate with an unfamiliar object in another different security domain. The establishment of a basic trust relationship for the two physical objects that do not know each other is the foundation of their security and privacy (Yao et al., 2020). Furthermore, Sharma et al. (2020) point out the shift of IoT to smart, connected and mobile IoT (M-IoT) devices. With this, they describe extended security, privacy and trust concerns for such networks and 'insufficient enforcement of these requirements introduces non-negligible threats to M-IoT devices and platforms' (Sharma et al., 2020, p. 1).

Again, despite this work, Ogonji et al. (2020, p. 1) highlight whilst IoT is still a growing and expanding platform, the current research in privacy and security shows there is 'little integration and unification of security and privacy that may affect user adoption of the technology because of fear of personal data exposure'. In the book *IoT: Security and Privacy Paradigm*, Pal et al. (2020) focus on bringing all security and privacy-related technologies into one source, so that students, researchers and practitioners can refer to this book for easy understanding of IoT security and privacy issues. This management of information into one resource is definitely a worthwhile endeavour and a way forward to understanding the wider IoT system and challenges at hand. Especially, as very recent research (Tian et al., 2020) concludes the security and privacy of IoT still faces a major challenge.

3 Studies

3.1 Study 1: Investigating People's Perceptions of Their Online Experience

The goal of this study 1 is to investigate the perceptions of people's online experiences (particularly around personal data, ownership, trust, accountability, privacy and security). It took place at Cardiff Met University in Autumn 2020 and aims to give insight into individuals' understanding and feelings around their data privacy whilst using online digital technologies. The study took approximately fifteen/twenty minutes in duration and allowed for statistical as well as qualitative analyses. The study was approved by the Ethics Board of School of Technologies, Cardiffmet and subjects provided online consent for study participation and the academic use of de-identified data.

Participant Demographics One hundred and thirteen participants completed the study. 46% of the participants were male, 53% female and 1% chose 'other'. 52.08% were aged 17–25, 20% aged 26–35, 14.58% aged 36–45, 7.29% between 46 and 55 and the remainder over 56 years of age. With regard to their educational background, 21.88% had a bachelor's degree, 13.54% a master's and 6.25% a PhD. With regard to IT proficiency, 58.24% were 'extremely comfortable' using technologies in computers and smart phones and 26.37% 'moderately comfortable'. 15.79% of all participants spend an average of less than 2 hours a week on social media, 37.89%

spend more than 2 hours a week but less than 2 hours a day, 28.42% spend 2–4 hours a day and 17.89% spend more than 4 hours a day.

Data Collection The study was conducted using the Qualtrics online survey software. Participants were presented with a series of quantitative and qualitative questions. In detail, some of these questions around ownership included: *How much music/movies/books do you download in a typical week?* and *To what extent, do you feel you own music/movies/books you download or stream to a device?* They were also asked about their feelings on: *In 2009, Amazon withdrew the book 1984 from Kindle. This affected users who had already paid and downloaded it. How do you feel about that?* and *If a bookstore had requested users return already paid for books and not offered a refund, how fair would you rate this move?*

Moreover, the study collected data on participants' feelings around trust: *Do you trust search engines such as Google to keep your data safe?* Participants were asked: *In your opinion, do you feel that your data is safe when you see a data protection cookie blocker?* and *In your opinion, do you feel that the big Five (Google, Amazon, Facebook, Netflix and Apple) handle your data responsibly?*

The study also collected participants' impressions around the access and sharing of their data: *How comfortable are you with governmental surveillance programmes that can access your personal text messages for the purposes of national security?* Participants were asked: *How comfortable are you with social media giants accessing your data on their product, for the purposes of offering you a better service?* and *How comfortable are you with social media giants accessing your data outside their product, for the purposes of offering you a better service? For example, there have been reports that Facebook traces users after they have logged off.* Finally, participants were asked: *You are filling out a job application and your prospective employer requires full access to your social media posts, how likely are you to comply?.* Participants were also asked: *You are filling out a job application and your prospective employer requires full access to your personal text messages, how likely are you to comply?* and *In general, how do you feel about your privacy when you go online?* For the statistical analysis, we focused on issues of ownership and trust. The results are detailed below.

Statistical Analysis As expected, there was correlation with how many movies, music and books were downloaded/streamed with how much time the participants spent on social media. There was no difference between 'ownership feel' in streaming or downloading.

In 2009, Amazon withdrew the book 1984 from Kindle. This affected the users who had already paid and downloaded it. We asked users how fair they found this move and we asked the same question if a physical book store had demanded the book back. The options were:

- It was extremely unfair/unjustifiable
- It was somewhat unfair/unjustifiable
- Neutral/not sure
- It was somewhat their choice/fair/justifiable
- It was fully their choice/fair/justifiable

An unexpected finding was that whilst most participants claimed they feel they 'somewhat owned' movies, books and music they downloaded or streamed onto their phones and computers, when asked about the controversial Amazon story, a small number (6%) of the same participants said they thought it was a 'fair move', yet had the same move been done in the real world they would find it unfair. This was a small number of participants, but it highlights the difference in how the virtual world is perceived. This was also correlated with age, as all those under 35 found the move unfair in both instances. As would have been expected, those who felt they did not own the material on their electronic devices were comfortable to have companies remove them at will.

We also conducted an ANOVA with the real-world behaviour as a baseline and social media and online behaviour as an 'intervention' to investigate how that affects people's behaviour. The scale used was:

- (−1) No ownership
- (0) Neutral/unsure
- (1) I somewhat own it
- (2) I own it to a large extent
- (3) I have full ownership and can copy it to other devices

The difference between music and movies is virtually none, and hence we only report one of them, there was however a difference for books, and the statistical tests are reported below: Fig. 1 is the t-test result of music downloaded with music bought in the form of a physical artefact. Figure 2 is the t-test result of music streamed vs music bought in the form of a physical artefact. In both the previous cases t Stat $< -t$ Critical two-tail so we reject the null hypothesis, and we observe that the difference between the sample means on both cases above convincing enough to suggest a significant difference in expectations of ownership online and in the physical word. That is not the case with books as we can see from the results below.

	Variable 1	Variable 2
Mean	0.462962963	2.1203703704
Variance	1.7836621668	0.1255624784
Observations	108	108
Pearson Correlation	0.6908252144	
Observed Mean Difference	-1.657407407	
Variance of the Differences	1.2553651783	
df	107	
t Stat	-15.37291105	
P (T<=t) one-tail	3.76888E-29	
t Critical one-tail	1.6592193119	
P (T<=t) two-tail	7.53776E-29	
t Critical two-tail	1.9823833702	

Fig. 1 The *t*-test result of music downloaded with music bought in the form of a physical artefact

	Variable 1	Variable 2
Mean	0.25	2.1203703704
Variance	1.4228971963	0.1255624784
Observations	108	108
Pearson Correlation	0.7020099222	
Observed Mean Difference	-1.87037037	
Variance of the Differences	0.9550017307	
df	107	
t Stat	-19.89012042	
P (T<=t) one-tail	4.919963E-38	
t Critical one-tail	1.6592193119	
P (T<=t) two-tail	9.839926E-38	
t Critical two-tail	1.9823833702	

Fig. 2 The *t*-test result of music streamed vs music bought in the form of a physical artefact

	Variable 1	Variable 2
Mean	1.9907407407	2.1203703704
Variance	0.383091035	0.1255624784
Observations	108	108
Pearson Correlation	0.6017018911	
Observed Mean Difference	-0.12962963	
Variance of the Differences	0.2447213569	
df	107	
t Stat	-2.723204286	
P (T<=t) one-tail	0.003775666	
t Critical one-tail	1.6592193119	
P (T<=t) two-tail	0.007551332	
t Critical two-tail	1.9823833702	

Fig. 3 The *t*-test result of books downloaded vs books bought physically

One interesting observation from the ownership questions is that although 85.26% of participants stated they did not download movies (and 31.58% did not stream), 40.13% scored on the scale of 1–3 for ownership of downloaded movies, against 20.36% for ownership of streamed movies. Figure 3 is the *t*-test result of books downloaded vs books bought physically.

When asked whether they trusted the Big Five (Google, Amazon, Facebook, Netflix and Apple) to handle their data responsibly, there was no significant difference between those who use social media more frequently and those who do not, for example, 27.91% of those who use social media 4 or more hours a day stated that they 'somewhat trusted' them, and the same modality of trust was found in 19.67% of those who use it less than 2 hours a day. Moreover, *t*-test comparisons between those who use social media less than 2 hours a day, between 2 and 4 hours a day and over 4 hours a day showed no statistically significant discrepancies. This

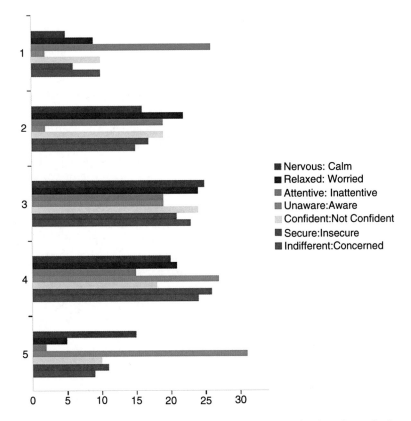

Fig. 4 How participants feel about being on social media with regard to the safety and privacy of their data

suggests that trust in data handling plays no effect in those who choose to opt out of social media completely or choose to moderate their usage. Figure 4 shows how participants feel about being on social media with regard to the safety and privacy of their data. When asked whether people should care more about privacy when online 45.68% strongly agreed and 29.63% agreed, whereas only 1.23% disagreed and 1.23% strongly disagreed.

To try and understand how participants might share data and trust their data to be handled properly, we asked them 'In your opinion, do you feel that your data is safe when you see a data protection cookie blocker?' and analysed the responses against how they trust in search engines to keep their data safe. We found that the two were strongly correlated as a Spearman correlation test returned a coefficient of approximately 0.51 showing them to be strongly correlated (Bobko, 2012; Bonett, 2008; Chen & Popovich, 2011).

Thematic Analysis For the qualitative analysis, we focused on issues of privacy, security and accountability. We have applied a thematic analysis to the data retrieved from the questionnaire. Firstly, the answers were read and re-read to enable the researchers to get an impression for what the participants were feeling. For example, words that were appearing most frequently in the dataset included privacy (22 times), data (18 times), need (18 times), care (10 times), information (10 times), aware (6 times) and consequences (4 times), to name a few. From these, themes such as 'education, governance, safeguards, awareness, care, accessibility, honesty, nothing will work, no idea, consequences, financial loss, no technology, research, evidence and caution' started to emerge. Time was then taken to gather all data relevant to each potential theme. Finally, a period of reviewing and refinement was undertaken and the following themes—education, governance, technical safeguard, awareness, caution, consequences and no technology—were determined to best demonstrate what participants believe needs to happen for them (and people in general) to care more about their privacy online.

As participant (2) highlighted, 'Educate users. It is a trade-off between level of privacy and level of service expected. Users should be educated enough to factor in the risks and use the services and expose their data as appropriate'. Another participant (1) spoke about the need for 'Far simpler information that is easily accessible for the public. Better governance of corporate behaviour, whilst maintaining a free Web'. In line with this, technical safeguards were another entity that was noted as being important: 'Antivirus and anti-cookies programs should be a part of the package when you buy a computer, not something you have to pay again and again after the purchase. No car is sold without a security belt or anti-crash in-built structures. Same should apply here' (participant (16)). Overall, it was felt that 'More people need to be made aware that really there is no privacy on the internet' (participant (18)) and that in general, people need to be more cautious 'People just need to make sure that they take care of their data' (participant (89)).

Moreover, the data quite strongly showed that exposing the consequences would be an effective motivation for people to care more about their privacy online. One participant (105) noted: 'I think it would require an invasion of someone's privacy to make them care more about their online privacy. Like someone who has their identity stolen or their credit card used without their permission online may be more concerned'. Whilst another stated: 'Something bad would need to happen for me to actually start caring. I currently do not understand the consequences of what my current privacy online is' (participant (103)). And again, 'I think that something bad would actually need to happen to me, like actually experience getting hacked rather than just hear other people's experiences' (participant (63)). A few participants had no idea how to solve this problem; however, there was one participant (111) who strongly felt: '...I believe we need to move further away from technology if we want to protect our privacy'.

The overall analysis of these answers indicates to us that there is a range of suggestions being offered however, exposure to the consequences, and seeing 'the dangers of what can happen to their data' (participant (108)) is one theme that was strongly highlighted as needing to happen in order for the test participant (and people in general) to care more about their privacy online.

Conclusion This study highlighted the compromises between privacy and security that people make on social media platforms and other IoT connected online venues. It presented statistically significant data to suggest a strong correlation between willingness to share data and trust in it being handled properly. It also presented results on data ownership and perceived rights as well as a glimpse at how people are feeling around accountability online. Emerging themes such as 'education, governance, technical safeguard, awareness, caution, consequences and no technology' highlight that people currently feel the need for more assurance and support as they interact and exist online. The main concern lies in the unrelenting growth of this need or more so, the lack of response or urgency to address these needs. As the IoT ecosystem opens up more and more opportunities for online communication and interaction, innovations emerge. In line with this, the need of society for clarity on data ownership, trust, accountability and privacy and society will exponentially grow too.

3.2 Study 2: The Digital Community's Perception to the Security in IoT

Social networks can provide a deep insight into people's interaction online. They can be an efficient method to extract the views of different individuals to understand how a community reacts to a certain topic (Agrawal et al., 2014). For this study, we started a Twitter collection service around the keywords *IoT security*, *IoT data* and *IoT privacy* with no location restrictions. In one week, we collected around 50K tweets.

Preprocessing the Data In detail, this study explores people's digital reaction to the concept of security in IoT. As the text is the dominant natural format of social networks (i.e. Twitter), the first step is to preprocess the data and then clean the text.

- Tokenizing. In natural language processing, one of the earliest steps during processing of the text is Tokenization (Grefenstette, 1999). It simply means dividing the tweet into one lower case version of the string where it feds to the following processing step.
- Removal of *Stop Words* and *Punctuation*. Stop words are a vital part of the natural language. It is common to remove the stop words and punctuation in the preprocessing phase because they create noise around the text, making extracting insightful information out of the text more difficult. In general, the text and the social network are articles, prepositions and pronouns. These do not give/add to the meaning of the sentence or the document (Vijayarani et al., 2015). Some examples for stop words include *the, in, a, an, with*. It is similar for punctuation (HaCohen-Kerner et al., 2020), and the removal of punctuation can improve the quality of the text and therefore increase the accuracy for the next steps in the process.

– Lemmatize. In the language we speak and write and more often in social networks, we use words that derived from another word, for example, (*running*, *runs*, they have one root *run*). The lemmatization is to return the word to the original/root form. The lemmatization proof needs to be very efficient in the natural language process (NLP) in order to improve the quality of the dataset and in doing so to prevent the redundancy in functionality and reduce the noise (Pradha et al., 2019; HaCohen-Kerner et al., 2020).

Exploring the Data In most NLP tasks, it is important to start the analysis by exploring the overview of the dataset. As explained earlier, the dataset used in this study consisted of 50K tweets around the #IoT. The *WordCloud* represents the frequency or the importance of each word, the bigger the word in the figure reflects the greater occurrence of it in the text. Figure 5 shows the word cloud that was generated after combining the full text of the 50K tweets. It is clear from this word cloud that the words *security*, *cyber security*, *privacy* and *artificial intelligent* (AI) stand out straight away. This highlights how important these words are with respect to the subject of *IoT* (Heimerl et al., 2014).

Fig. 5 Word cloud in Twitter around IoT

The word cloud certainly gives an excellent context to the text. Similarly, another approach is to explore the emotions raised in the text. For many years, the detection of emotions in the text has been a hot area of research for the field of NLP (Seyeditabari et al., 2018).

Furthermore, the tweets are associated with hashtags, which are essentially labels for the content. These hashtags help to gather similar content (i.e. bring similar content together under the same hashtag). This makes it easier for others to find content which discusses the same topic hashtags; these are commonly used in social networks. Although Twitter initially introduced hashtags, they have been adopted by other social networks (i.e. Facebook and Instagram) on their social media sites (Kouloumpis et al., 2011). Using a hashtag can be an effective way to quickly contextualize what the topic is about without using up valuable characters or writing repetitive captions. Therefore, it is essential to extract the hashtags to add a context to the tweets (Bruns & Stieglitz, 2013).

Figure 6 reflects on the top 10 hashtags around the topic of *IoT*. These match the findings from the word cloud around *#cybersecurity*, *#security* and *#AI* and then show a big difference between the remaining hashtags. Moreover, there are other popular handlers in the social network world, such as @ which used to mention a certain user. Figure 7 shows the top 10 users who were mentioned in the context of IoT. The top three users were all experts in the field of *data science*, which shows the high correlation between the two domains (i.e. data science and IOT). Also, it is

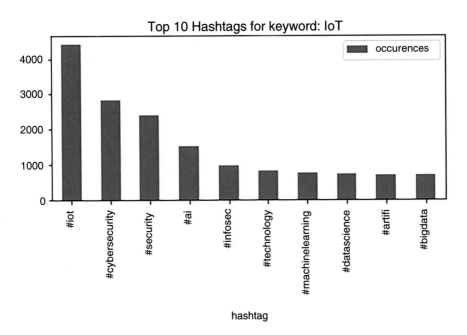

Fig. 6 Top 10 hashtags—IoT

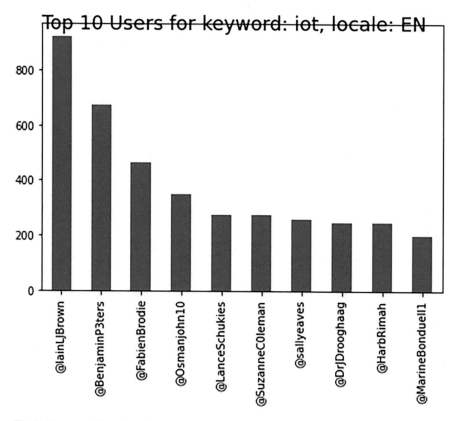

Fig. 7 Top users @mentioned—IoT

worth mentioning the fourth top user is a researcher in *cybersecurity*, which again highlights the association of *IoT* and *cybersecurity*.

In terms of the affective, there are commonly two approaches to extract emotions from text, *Machine learning* (ML) and *Lexicon approach*. There are plenty of methods within the *machine learning* approach to classify the text and then associate an emotion to the text. However, in this study, we are using the *Lexicon approach* (Li & Xu, 2014), as it has been proven to have a more effective output with regard to tweets, especially as the nature of the tweets has a limited number of characters. We have applied a widely used library called NRC Lexicon (Mohammad & Turney, 2013), which is based on the NRC dictionary to extract the frequency of the emotions. The eight common basic emotions are defined as *trust, anticipation, anger, fear, surprise*, sadness, disgust and *joy* (Izard, 1992).

Figure 8 shows the overall emotions; the distribution of the emotions in the text highlights that *trust, anticipation, fear* and *joy* have a higher percentage of domination 28.4%, 18.9%, 17.9% and 16.2%, respectively. On the other hand, the

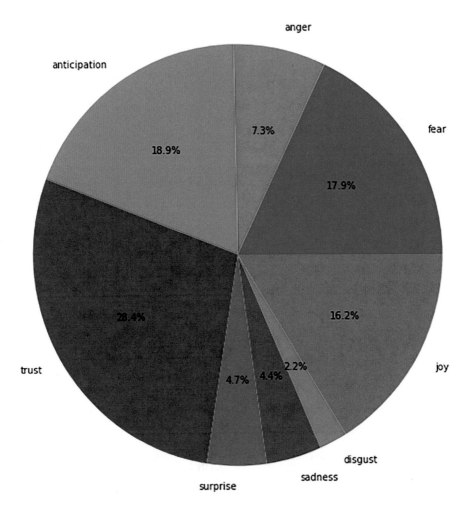

Fig. 8 Overall emotions in Twitter around IoT

emotions *disgust*, *sadness* and *surprise* are the lowest with 2.2%, 4.4% and 4.7%, respectively.

Further drilling into the emotions, Fig. 8 presents a breakdown to understand the context of each emotion. A word cloud per emotion demonstrates the frequency of words within each emotion (see the breakdown in Fig. 9). It is clear that the *security* keyword is a buzz word around the topic *IoT* except for the emotions *joy* and *disgust*. However, for the other emotions, there is also an association with other keywords, for example, in anger the buzz words associated with it are *machine learning*, *privacy*, *hacking*, *data breach* and *data security*. Trust was also one of the top

Fig. 9 Word cloud for each individual emotion

emotions raised in the overall analysis. Further drilling of the keywords highlights major buzz words such as *intelligence, block chain, exploiting* and *improve.*

Applying Topic Modelling In this section, we apply topic modelling to the collected dataset to add more context to the analysis. In NLP, topic detection is the function of grouping words in a corpus of text (Onan et al., 2016). One of the common applications for topic modelling is to derive useful content from unstructured text (i.e. Twitter). The approach we follow for the topic modelling is latent Dirichlet allocation (LDA). LDA is a Bayesian Hierarchy model, in which a corpus of text is divided into entities namely *words, documents* and *corpora. Word* defines a keyword, and it is the smallest unit in the document, and it builds the document. A *document* is a group of N words. *Corpus* is a collection of documents. However, topics are the vocabulary distribution, and each document contains a certain distribution of the topics (Negara et al., 2019).

In this study, we apply topic modelling (LDA) to our collection of tweets to extract the topics. It is important to note that LDA initially does not label the topic. It groups relevant words to build a topic. In Fig. 10, the topics from our tweets were distributed as shown. From the structure of the keywords, we can assume a label for each topic as follows:

– Topic 0 = AI and Data Science
– Topic 1 = Privacy
– Topic 2 = Cyber Security
– Topic 3 = Future of IoT

In addition, we attempt to explain how emotions are contributing in each topic, by diving into the topic and extracting emotions per topic as seen in Fig. 11. In topic 0 (AI and data science), the dominant emotions are *fear, trust, anticipation* and *joy,* in topic 1 (privacy), emotions are *fear, trust, anticipation* and *anger,* in topic 2 (cyber security), emotions are *anticipation* and *trust* and in topic 3 (future of IoT), emotions are slight of *fear, joy* and high value of *anticipation* and *trust.*

Conclusion Although cybersecurity is a big part of the conversation, as the previous analysis revealed in Figs. 8 and 9, the most dominant emotion is *trust* with 28.4%. The numbers indicate that although the online community is fully aware of the potential risks of *IoT* (i.e. cyber security, phishing and data breach), the vast majority believe in the science and feel that *trust* can overcome these obstacles. Interestingly, the second top major emotion was *fear* (17.9%), which can be understood as the *IoT* ecosystem is now touching everyone's daily routines/lives and homes. On the other hand, the third top emotion was *joy* (16.2%), which also reflects that these technologies and devices are agreeable and accepted with some precautions. All these meanings and emotions are strongly evident in the data.

Notably, anger is associated with privacy, security and hack, which is a big concern for the digital community. Furthermore, *fear* features similar concerns but with a smaller magnitude than the overall emotions. The trust indicates that the technology of data science and machine learning can overcome these issues.

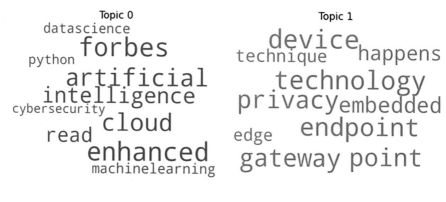

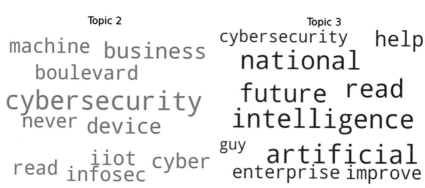

Fig. 10 Applying LDA topic modelling to the IoT tweets

The output of the LDA concluded the topics could label into four main topics: (1) AI and Data Science, (2) Privacy, (3) Cyber Security and (4) Future of IoT. As we have discussed, LDA (topic modelling) was applied to reveal these topics and show the distribution of the topic among the full dataset. Figure 11 explained the emotions among each individual topic. In the topic *AI and data science*, the overall mean of emotion is 0.16, and the four emotions dominating this topic are fear, anticipation, trust and joy. Despite the four being very close in values, the data does show the contradictions in the thoughts and ideas around *IoT* as a domain and the association of the data science and AI to solve and ease the fear. Furthermore, the topic *Future of IoT* is very similarly distributed to that of previous topic but with a wider gap between the *trust* and the *fear* and *anticipation* and *joy*. These findings confirm that the digital community trusts that the future is positive, and they are fully aware of the particular issues and risks. On the other hand, the topics *privacy* and *cyber security* enjoy the same pattern of emotions, except for *anger*. The topic of *privacy* is higher for *anger* than the topic of *cyber security*. A reasonable explanation would be that humans are emotively triggered when their privacy is invaded, and this will raise their anger more than the fear emotion in both topics.

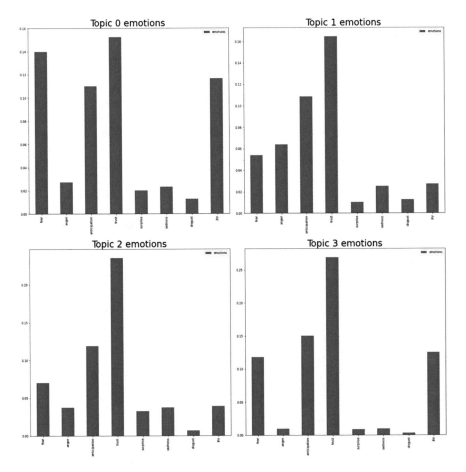

Fig. 11 Extract emotions by topics

4   Discussion

Statistical results on data ownership and perceived rights show correlations with how movies, music and books downloaded/streamed with how much time the participants spent on social media. There were, however, no differences between 'ownership feel' in streaming and downloading. With regard to ownership of content downloaded or bought in the physical world, our t-test analyses showed that the difference between the sample means on both cases was convincing enough to suggest a significant difference in expectations of ownership online and in the physical word (on all content analysed). With regard to frequency of usage, we noted no significant difference between those who use social media more frequently and those who do not in terms of trusting tech giants with regard to the safety and privacy of their data. Finally, we found that trust and willingness to share data were strongly correlated as a Spearman correlation test returned a coefficient of approximately 0.51 showing them to be strongly correlated.

Humans are hardwired to trust one another, and they have a social instinct to get along with other people. With the rise of IoT (and the new means of communication and interaction it affords), we have seen an intensification of this human need for trust, approval and to be admired. We have also seen an increase of opportunities to manipulate trust, to spread misinformation and disinformation, to influence public opinion and exercise control, to undermine and divide societies etc. To the extent, that it is having huge implications not only on people's privacy and security but also on their self-respect and self-worth. The more devices connected, the more data shared and potentially the more revenue generated. In the midst of this, people are resigned to giving up most of their privacy for the stamp of social approval, sense of personal value and convenience that IoT applications can enable and afford (Mineo, 2017). To add further to this complex situation, in our study 1, we have found that 45.68% of participants strongly agreed and 29.63% agreed that people should care more about privacy when online. The analysis highlights themes such as education, governance, technical safeguard, awareness, caution, consequences and no technology to best demonstrate what participants believe needs to happen for them (and people in general) to care more about their privacy online. The findings show thoughts about better governance and technical securities and exposing the consequences as an effective motivation for people to care more about their privacy online.

Furthermore, this chapter discussed the reflections (feelings) of people on IoT (privacy and security) on Twitter. Indeed, nowadays, social networks play a significant role in people's daily lives, and social network analysis is a successful way to explore people's behaviours and their pattern of relationships (i.e. it can reflect topics, trends, services and domains). As we have seen, study 2 collected tweets with the keyword *IoT* which had no location restriction. This investigation of 50K tweets shows a clear indication that security and cybersecurity are central to the conversation around *IoT* (see Fig. 5). In addition, other buzz words such as artificial intelligence and machine learning are noted in these conversations. These findings highlight the importance of artificial intelligence in the *IoT* domain. It is not hugely surprising as AI nowadays plays a core role in most research areas, which reflect in Fig. 6. Moreover, the top ten hashtags highlight the association of AI and machine learning with cybersecurity, representing how AI and ML use tools to combat the cybersecurity threats in the IoT world.

The analysis also indicates that in a conversation around IoT, there is an explicit mention of cybersecurity, AI and data science experts, again reinforcing the previous findings to highlight how these three domains are tightly interlinked. Overall, the analysis shows that people are fully aware of IoT risks yet are ready to combat these risks with science and technology. Furthermore, the reflection on the future of the IoT is positive and promising despite these risks and concerns. It is evident from the findings that the risks and concerns do not detract from or eliminate the happiness gained from the convenience achieved by using these IoT technologies for daily life.

5 Conclusion

Big tech companies have created empires from people's data; their likes, dislikes, photos and comments have generated them billions in revenue. It seems that the everyday person has become the product they sell, yet these people (us) have no control over how or what exactly they sell, we are blocked from even seeing our data, and the profiles they build from our activities. The concern here is that with IoT being deployed everywhere and billions of devices connected and being used, these existing problems around ownership, trust, accountability and privacy and security will be fully exacerbated.

Indeed, for decades, society has blindly used products, trusting some faceless corporation to somehow have decent codes of ethics in how they use our data; the situation is no longer the same. As people acquire and implement more and more interconnected IoT devices, the risks to privacy and security increase. The more devices on the network, the more vulnerabilities that are created and the more data that flow. It goes without saying that data ownership (particularly IoT data ownership) gets confusing, accountability is hard to pin down and trust goes out the window.

Following several documentaries and journalistic exposes, the personal privacy and security situation is becoming increasingly unnerving to anyone paying attention. Among the most concerning issues are now well documented, interference in elections, the ability to sway large numbers to vote for a candidate or particular political party. But this was not an issue born overnight; we slowly allowed these corporations to encroach on our privacy and our data rights in exchange for the convenience of their (seemingly free) technology. As we have seen in this chapter, IoT brings with it significant risks to our security and privacy; however, the convenience and benefits of connectivity continue to outweigh the risks and attract users. It has, in fact, come at a very high price. This has been highlighted in the academic community (Determann, 2012; Madden, 2012; Smith et al., 2012).

In parallel, due to the rise of IoT and the exponential increase of available information, we are seeing more and more opportunities for misinformation and fake news. Not only are we essentially watching our democracies get high-jacked for profit, but also the spread of misinformation has come at deadly consequences during the on-going Covid pandemic (Roozenbeek et al., 2020; Tasnim et al., 2020). Academics, scientists and medical practitioners have attempted to combat the spread of dangerous untruths, to direct people to verifying sources before listening to what is being said or written, but it is very difficult for these posts to combat the vast seas of conspiracy theories created through the vast network of IoT devices.

Social media giants have made public claims and seem to have ramped up the effort to combat misinformation, and however there's very little legal incentive for them to do so. One of the most important Internet laws lies in Section 230 of the Communications Decency Act, whilst giving giants like Facebook, Twitter, YouTube the right to moderate content, it also protects these companies from being held responsible for what is posted on their sites. This law was written long before these companies and the world of IoT was born and in a world where Netscape, AOL and mnbjhhn dominated. These are very different from what we are using today. Similar laws exist in democracies throughout the world, for example, the Defamation Act 2013 passed in the United Kingdom in a way removes responsibility to those hosting offensive content, if they can prove that they did not post the offence. This provides immunity to social media giants for cases where defamatory statements are posted on their platforms.

In this chapter, we have presented people's perception of their own data rights online and how it differs from the external rights they enjoy in the real world. The authors have also provided a snapshot of how people are currently feeling about IoT and its impact on their privacy and security. The findings have clearly shown that people are willing to oversee the risks of IoT for the convenience that the IoT ecosystem affords. Yet, the authors have also shown the many values that people expect to maintain (around ownership, trust, accountability, privacy and security) when crossing over to their virtual presence. The question now lies in how we continue to move forward to design an IoT infrastructure that mitigates the risks to human well-being yet enables and ensures these desired values?

References

Agrawal, D., Budak, C., El Abbadi, A., Georgiou, T., & Yan, X. (2014). Big data in online social networks: User interaction analysis to model user behavior in social networks. In A. Madaan, S. Kikuchi, & S. Bhalla (Eds.), *Databases in networked information systems* (pp. 1–16). Cham: Springer International Publishing.

Alfandi, O., Khanji, S., Ahmad, L., & Khattak, A. (2021). A survey on boosting IoT security and privacy through blockchain: Exploration, requirements, and open issues. *Cluster Computing, 24*(1), 37–55.

AlHogail, A. (2018). Improving IoT technology adoption through improving consumer trust. *Technologies, 6*(3), 64.

Alraja, M. N., Farooque, M. M. J., & Khashab, B. (2019). The effect of security, privacy, familiarity, and trust on users' attitudes toward the use of the IoT-based healthcare: The mediation role of risk perception. *IEEE Access, 7*, 111341–111354.

Alshohoumi, F., Sarrab, M., AlHamadani, A., & Al-Abri, D. (2019). Systematic review of existing IoT architectures security and privacy issues and concerns. *International Journal of Advanced Computer Science and Applications, 10*(7), 232–251.

Altaf, A., Abbas, H., Iqbal, F., & Derhab, A. (2019). Trust models of Internet of smart things: A survey, open issues, and future directions. *Journal of Network and Computer Applications, 137*, 93–111.

Asplund, M., & Nadjm-Tehrani, S. (2016). Attitudes and perceptions of IoT security in critical societal services. *IEEE Access, 4*, 2130–2138.

Atlam, H. F., & Wills, G. B. (2020). IoT security, privacy, safety and ethics. In *Internet of Things*.

Barakati, B. A., & Almagwashi, H. (2020). IoT of trust: Toward ownership management by using blockchain. *International Journal of Computer Science and Information Security (IJCSIS), 18*(3).

Bobko, P. (2012). *Correlation and regression: Applications for industrial organizational psychology and management*. Sage

Bonett, D. G. (2008). Meta-analytic interval estimation for bivariate correlations. *Psychological Methods, 13*(3), 173.

Bruns, A., & Stieglitz, S. (2013). Towards more systematic twitter analysis: Metrics for tweeting activities. *International Journal of Social Research Methodology, 16*(2), 91–108.

Caminha, J., Perkusich, A., & Perkusich, M. (2018). A smart trust management method to detect on-off attacks in the internet of things. *Security and Communication Networks.* https://doi.org/10.1155/2018/6063456

Chen, P. Y., & Popovich, P. M. (2011). *Correlation: Parametric and nonparametric measures.* Sage.

Ciardiello, F., & Di Liddo, A. (2020). Privacy accountability and penalties for IoT firms. *Risk Analysis.* https://doi.org/10.1111/risa.13661

Crabtree, A., Lodge, T., Colley, J., Greenhalgh, C., Glover, K., Haddadi, H., Amar, Y., Mortier, R., Li, Q., Moore, J., Wang, L., Yadav, P., Zhao, J., Brown, A., Urquhart, L., & McAuley, D. (2018). Building accountability into the Internet of Things: The IoT Databox model. *Journal of Reliable Intelligent Environments, 4*(1), 39–55.

Determann, L. (2012). Social media privacy: A dozen myths and facts. *Stanford Technology Law Review, 7*, 1–14.

Economides, A. A. (2017). User perceptions of Internet of Things (IoT) systems. In *Communications in Computer and Information Science.*

Emami-Naeini, P., Agarwal, Y., Faith Cranor, L., & Hibshi, H. (2020). Ask the experts: What should be on an IoT privacy and security label? In *Proceedings - IEEE Symposium on Security and Privacy.*

Grefenstette, G. (1999). Tokenization. In *Syntactic Wordclass Tagging* (pp. 117–133). Springer.

HaCohen-Kerner, Y., Miller, D., & Yigal, Y. (2020). The influence of preprocessing on text classification using a bag-of-words representation. *PLoS ONE 15*(5), e0232525.

Heimerl, F., Lohmann, S., Lange, S., & Ertl, T. (2014). Word cloud explorer: Text analytics based on word clouds. In *2014 47th Hawaii International Conference on System Sciences* (pp. 1833–1842). IEEE.

Hudson, F. D. (2018). Enabling trust and security: TIPPSS for IoT. *IT Professional, 20*(2), 15–18.

IS. (2019). The Trust Opportunity: Exploring Consumer Attitudes to the Internet of Things. Internet Society. https://www.internetsociety.org/resources/doc/2019/trust-opportunity-exploring-consumer-attitudes-to-iot/

Izard, C. E. (1992). Basic emotions, relations among emotions, and emotion-cognition relations.

Jakobi, T., Stevens, G., Castelli, N., Ogonowski, C., Schaub, F., Vindice, N., Randall, D., Tolmie, P., & Wulf, V. (2018) Evolving needs in IoT control and accountability. In *Proceedings of the ACM on Interactive, Mobile, Wearable and Ubiquitous Technologies.*

Kotis, K., Athanasakis, I., & Vouros, G. A. (2018). Semantically enabling IoT trust to ensure and secure deployment of IoT entities. *International Journal of Internet of Things and Cyber-Assurance, 1*(1), 3–21.

Kouloumpis, E., Wilson, T., & Moore, J. (2011). Twitter sentiment analysis: The good the bad and the OMG! In *Proceedings of the International AAAI Conference on Web and Social Media* (Vol. 5).

Lessig, L. (2006). IOT. Codev2 - LESSIG. https://lessig.org/product/codev2. April 04, 2021.

Li, W., & Xu, H. (2014). Text-based emotion classification using emotion cause extraction. *Expert Systems with Applications, 41*(4), 1742–1749.

Madden, M. (2012). Privacy management on social media sites. *Pew Internet Report, 24*, 1–20.

Mashhadi, A., Kawsar, F., & Acer, U. G. (2014). Human data interaction in IoT: The ownership aspect. In *2014 IEEE World Forum on Internet of Things, WF-IoT 2014.*

Meng, W. (2018). Intrusion detection in the Era of IoT: Building trust via traffic filtering and sampling. *Computer, 51*(7), 36–43.

Mineo, L. (2017). When it comes to Internet privacy, be very afraid, analyst suggests – Harvard Gazette. https://news.harvard.edu/gazette/story/2017/08/when-it-comes-to-internet-privacy-be-very-afraid-analyst-suggests/

Mohammad, S. M., & Turney, P. D. (2013). NRC emotion lexicon. National Research Council, Canada 2.

Mohammadi, V., Rahmani, A. M., Darwesh, A. M., & Sahafi, A. (2019). Trust-based recommendation systems in Internet of Things: A systematic literature review. *Human-Centric Computing and Information Sciences, 9*(1), 1–61.

Negara, E. S., Triadi, D., & Andryani, R. (2019). Topic modelling twitter data with latent Dirichlet allocation method. In *2019 International Conference on Electrical Engineering and Computer Science (ICECOS)* (pp. 386–390). IEEE.

Ogonji, M. M., Okeyo, G., & Wafula, J. M. (2020). A survey on privacy and security of Internet of Things. *Computer Science Review, 38*, 100312.

Onan, A., Korukoglu, S., & Bulut, H. (2016). LDA-based topic modelling in text sentiment classification: An empirical analysis. *International Journal of Computational Linguistics and Applications, 7*(1), 101–119.

Online Trust Alliance. (2017). IoT Trust Framework. Online Trust Alliance.

Pal, S., García, D. V., & Le, D. N. (2020). *IoT: Security and privacy paradigm*. CRC Press.

Perzanowski, A., & Schultz, J. (2016). *The end of ownership: Personal property in the digital economy*. MIT Press.

Pradha, S., Halgamuge, M. N., & Vinh, N. T. Q. (2019). Effective text data preprocessing technique for sentiment analysis in social media data. In *2019 11th International Conference on Knowledge and Systems Engineering (KSE)* (pp. 1–8). IEEE.

Roozenbeek, J., Schneider, C. R., Dryhurst, S., Kerr, J., Freeman, A. L., Recchia, G., Van Der Bles, A. M., & Van Der Linden, S. (2020). Susceptibility to misinformation about COVID-19 around the world: Susceptibility to COVID misinformation. *Royal Society Open Science, 7*(10), 201199.

Rosenblatt, B. (2016). What Does Ownership Mean in the Digital Age? Copyright and Technology.

Seyeditabari, A., Tabari, N., & Zadrozny, W. (2018). Emotion detection in text: A review. Preprint, arXiv:1806.00674.

Sharma, V., You, I., Andersson, K., Palmieri, F., Rehmani, M. H., & Lim, J. (2020). Security, privacy and trust for smart mobile-Internet of Things (M-IoT): A survey. *IEEE Access, 8*, 167123–167163.

Sicari, S., Rizzardi, A., Grieco, L. A., & Coen-Porisini, A. (2015). Security, privacy and trust in Internet of Things: The road ahead. *Computer Networks, 76*, 146–164.

Singh, J., Millard, C., Reed, C., Cobbe, J., & Crowcroft, J. (2018). Accountability in the IoT: Systems, law, and ways forward. *Computer, 51*(7), 54–65.

Smith, M., Szongott, C., Henne, B., & Von Voigt, G. (2012). Big data privacy issues in public social media. In *IEEE International Conference on Digital Ecosystems and Technologies*.

Stallings, W. (2011). *Network security essentials: Applications and standards*. Pearson.

Tasnim, S., Hossain, M., & Mazumder, H. (2020). Impact of rumors and misinformation on COVID-19 in social media. *Journal of Preventive Medicine and Public Health, 53*(3), 171–174.

Tawalbeh, L., Muheidat, F., Tawalbeh, M., & Quwaider, M. (2020). IoT privacy and security: Challenges and solutions. *Applied Sciences (Switzerland), 10*(12), 4102.

Tewari, A., & Gupta, B. B. (2020). Security, privacy and trust of different layers in Internet-of-Things (IoTs) framework. *Future Generation Computer Systems, 108*, 909–920.

Tian, H., Ge, X., Wang, J., Li, C., & Pan, H. (2020). Research on distributed blockchain-based privacy-preserving and data security framework in IoT. *IET Communications, 14*(13), 2038–2047.

Tzafestas, S. (2018). Ethics and law in the internet of things world. *Smart Cities, 1*(1), 98–120.

Urquhart, L., Lodge, T., & Crabtree, A. (2019). Demonstrably doing accountability in the Internet of Things. *International Journal of Law and Information Technology, 27*(1), 1–27.

van Deursen, A. J. A. M., van der Zeeuw, A., de Boer, P., Jansen, G., & van Rompay, T. (2019). Digital inequalities in the Internet of Things: Differences in attitudes, material access, skills, and usage. *Information, Communication & Society, 24*(2), 258–276.

Vijayarani, S., Ilamathi, M. J., Nithya, M., et al. (2015). Preprocessing techniques for text mining-an overview. *International Journal of Computer Science & Communication Networks, 5*(1), 7–16.

Voas, J., Kuhn, R., Laplante, P., & Applebaum, S. (2018). *Internet of Things (IoT) trust concerns (draft).* NIST.

Williams, M., Nurse, J. R., & Creese, S. (2018). Privacy is the boring bit: User perceptions and behaviour in the Internet-of-Things. In *Proceedings - 2017 15th Annual Conference on Privacy, Security and Trust, PST 2017.*

Yadav, P., Safronov, V., & Mortier, R. (2019). Enforcing accountability in Smart built-in IoT environment using MUD. In *BuildSys 2019 - Proceedings of the 6th ACM International Conference on Systems for Energy-Efficient Buildings, Cities, and Transportation.*

Yan, Z., Zhang, P., & Vasilakos, A. V. (2014). A survey on trust management for Internet of Things. *Journal of Network and Computer Applications, 42,* 120–134.

Yao, X., Farha, F., Li, R., Psychoula, I., Chen, L., & Ning, H. (2020). Security and privacy issues of physical objects in the IoT: Challenges and opportunities. *Digital Communications and Networks, 7*(3), 373–384.

Zheng, S., Apthorpe, N., Chetty, M., & Feamster, N. (2018). User perceptions of smart home IoT privacy. *Proceedings of the ACM on Human-Computer Interaction, 2,* 1–20.

Zhou, W., Jia, Y., Peng, A., Zhang, Y., & Liu, P. (2019). The effect of IoT new features on security and privacy: New threats, existing solutions, and challenges yet to be solved. *IEEE Internet of Things Journal, 6*(2), 1606–1616.

Covid-19 Era: Trust, Privacy and Security

Vinden Wylde, Edmond Prakash, Chaminda Hewage, and Jon Platts

1 Introduction

The main purpose of this chapter is to highlight and demonstrate instances of how institutions, business and individuals use Big Data (BD) and other critical component technologies such as Artificial Intelligence (AI), Blockchain (BC), Edge and Cloud infrastructures, to interpret and apply strategic principles in facilitating innovation with BD and the General Data Protection Regulations (GDPRs) to include pandemic situations.

At the time of writing (see Fig. 1), the novel coronavirus disease 19 (Covid-19) has affected most, if not all of the globe to date (WHO Coronavirus Disease, 2021). Covid-19 is caused by severe acute respiratory syndrome coronavirus 2 (SARS-Cov-2), which itself has affected at least 214 countries and continues to spread with minimal sign of being in a well-controlled global situation (Tran & Ngoc, 2020). In March 2020, the World Health Organization (WHO) officially declared a novel coronavirus (2019-nCoV) or Covid-19 outbreak.

By the 11th of March, the WHO announced a Covid-19 pandemic that had spread worldwide (Covid-19 Alert, 2020). Due to the Covid-19 global pandemic, further questions are raised which bring significant emphasis on digital information quality and trust.

Since December 2019, the novel coronavirus strain has created an unprecedented outbreak which further changes and challenges the ability to predict the environment in which we all live.

Whether it is the privacy of a video conference call, the availability of basic products, the cumulative effect of co-worker behaviours, stock prices and job security,

V. Wylde · E. Prakash (✉) · C. Hewage · J. Platts
Cardiff School of Technologies, Cardiff Metropolitan University, Cardiff, UK
e-mail: vwylde@cardiffmet.ac.uk; eprakash@cardiffmet.ac.uk; chewage@cardiffmet.ac.uk; jplatts@cardiffmet.ac.uk

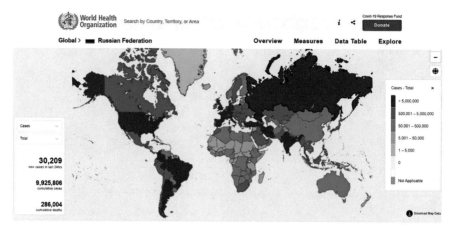

In Russian Federation, from 3 January 2020 to 4:08pm CET, 9 December 2021, there have been 9,925,806 confirmed cases of COVID-19 with 286,004 deaths, reported to WHO. As of 7 December 2021, a total of 130,634,456 vaccine doses have been administered.

Fig. 1　WHO coronavirus dashboard

the confidence of prediction in multiple domains appears to have diminished. This includes political and social challenges that have arisen from the UK's post-Brexit scheme.

For example, trust in government measures and implemented strategies regarding Covid-19 are under continual pressure to deliver strategies, interventions and outcomes that show the effect of prescribed legislation and guidance on Covid-19 mitigation (Living with COVID-19, 2020). BC technologies, for example, reinforce the traceability and infallibility of information; hence, BC is a proposed method in support of BD technologies.

For society to function, healthy levels of trust and decision-making for citizens, businesses and services are paramount. In turn, this further complicates and contributes to a significant change in world behaviour, cultures (i.e., mask wearing, daily hand-washing routines and local/international restrictions), societies and individuals. This is especially true for individuals in an ever-changing digital world whose trust depends upon criteria such as their expectations, control and vulnerabilities (agency) (Goddard, 2017).

Below are chapter themes and critical component technologies that symbiotically support efforts in Covid-19 resource planning, strategy and deployment.

- Coopetition, Information and Trust
- Big Data and Cloud Computing
- Internet of Things (IoT) and Edge Computing
- Blockchain and Communication Networks
- Legal and Ethical Challenges that conflict between Data Security, Data Protection and Ethics.

The chapter structure is as follows. Section 2 gives various accounts of previous works on topics such as Coopetition, GDPR, Big Data and Covid-19, with technological and legal digital innovations. Section 3 provides insights into concepts and strategies that encapsulate the delivery of a successful strategy during a pandemic or crisis situation. Section 4 demonstrates legal, ethical and technical challenges inherent to critical components and compliance. Section 5 presents findings and insights that may contribute and enhance knowledge and the wider public. Section 6 includes any additional policy, processes or system suggestions for future works.

2 Previous Work

2.1 Coopetition, Information and Trust

In 2007, K. Walley of Harper Adams University, UK, published an article examining coopetition. This term is utilised to encapsulate the cooperative and competitive nature of traditional inter-firm dynamics. Previous studies used terms such as 'inter-firm-cooperation' and alliances (James & Crick, 2020). Although coopetition has only found academic attention and traction in more recent years, evidence exists that firms and organisations have been undertaking such relationships and practices for a considerable length of time.

Fast-forward to today, and there is global recognition that firms cooperate and compete simultaneously, alongside a large body of academic works that influence subject areas such as, Strategic Management, Marketing and Supply Chain Management. However, Crick (2020) states that coopetition in principle has a positive overall effect on company performance, yet little is known on how the implementation of business-to-business marketing strategies bare under the significant systematic weight of a large-scale or global emergency (Medrano & Olarte-Pascual, 2016).

Schiffling (2020) examines coopetition in the context of swift trust and swift distrust in humanitarian operations. Their findings conclude from the case studies of 18 global humanitarian organisations, 48 interviews and analysis of public documents that coopetition is supported by both swift trust and swift distrust. A coopetitive strategy potentially enables actors and organisations (usually Non-Governmental Organizations (NGOs)) to deliver disaster relief to include the military, governments and local communities, whilst simultaneously competing for media attention and resources (Schiffling et al., 2020). In terms of trust, it is clear that the relationship between coopetition and trust engenders uncertainty, interdependence and carries general over-arching concerns for opportunism (Living with COVID-19, 2020).

An example in The Hindu Times reports that the Chief Justice of India (CJI) Sharad A. Bobde stated that Indians had grave apprehensions regarding privacy from Facebook and WhatsApp. Petitioners alleged that a new privacy policy had been introduced by WhatsApp which scraps a user's 'opt-out' policy for a 'compulsory consent to share data' with Facebook and affiliated groups policy (WhatsApp Policy, 2021). The defence stated that the policy did not discriminate and is applied globally except Europe, which has a special law called General Data Protection Regulations (GDPRs). The challenge was not about encryption of messages, but concerns over profiting from the sharing of meta-data and the clear difference of privacy rules between Europe and India.

2.2 General Data Protection Regulations

The GDPR legal framework entered force on 24 May 2016, applicable by 25 May 2018, that harmonises sets of rules that apply to all personal data that is processed within EU digital boundaries (see Fig. 2). The main objective of which is to ensure high standards of protection and enhancement of individual and organisational legal certainty, for example, the Police Data Protection Directive (PDPD) which monitors competent authorities in their processing of Personal Data (Zaeem & Barber, 2020). The GDPR and PDPD superseded Directive 95/46/EC on 27 April 2016, to facilitate the private and most of the public sectors and the Council Framework Decision 2008/977/JHA for the law enforcement sectors. Regulation 2018/1725, applicable by 11 December 2018, sets out a series of data protection rules for all EU institutions (Caruccio et al., 2020; Shastri, 2019).

This legal framework clarifies and supports Data Protection Officers (DPOs) objectives in carrying out their roles, defines clear mechanisms for the exercising of rights and provides accessibility and transparency from within EU institutions and of the European Data Protection Supervisor (EDPS).

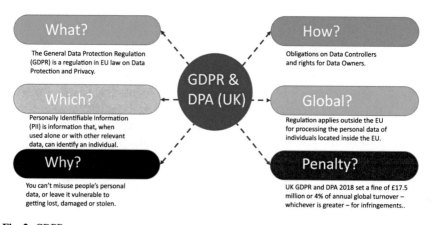

Fig. 2 GDPR

In this context, for GDPR to successfully operate and utilise all digital resources in an equitable manor, 6 core principles are specified by GDPR to help understand the main crux of GDPR.

- Lawfulness, fairness and transparency
- Purpose, storage limitation and accuracy
- Data minimisation, integrity and confidentiality (Goddard, 2017; Poplavska, 2020; Thorgren, 2019)

For purposes of prevention, investigation, detection and prosecution of criminal offences, specific mandates apply to all data controllers and processors that operate within the EU boundary to *implement appropriate technical and organisational measures [. . .] to ensure the ongoing confidentiality, integrity and availability and resilience of processing systems and services*, with heavy penalties including up to €20 million fines or 4% of global annual turnover, 'whichever is the greatest' (IT Governance Privacy Team, 2017).

As with regard to implementation and enforcement, there appears to be consensus amongst researchers that technological developments tend to out-pace legal frameworks (Shastri, 2019). The main challenge here is tackling the range of lesser Data Protection frameworks in operation globally, the protections of GDPR need to be robust enough to also combat under developed regulation and to remain equitable, transparent and fair to users. However, transparency and fairness tend to be perceived differently across the globe.

2.3   Contact Tracing

Across the globe, cloud and cell networks utilise smartphones and applications in disease risk mitigation and continue to draw controversy with countries seeking different implementation strategies in minimising disease transmission (Wylde et al., 2021). For example, Ant Financial (sister company to Alibaba) produced the application 'Alipay Health Code' that was deployed in Hang-zhou, China. Its algorithm collects and matches addresses, self-reporting of symptoms information and personal ID, whilst other meta-data is collected passively to include GPS location data. Then, colour codes (Red, Amber and Green) and Quick Response (QR) codes enable a given user to determine whether to access public areas to include malls, markets and subways.

In contrast, the Guardian reports concern regarding the UK government, specifically England's test-and-trace system (costs £22bn and in operation since 23 April 2020) with allegations of contact tracers working from abroad, and how one company responded with introducing the tracking of employee locations. The outsourcing firm Serco is believed to have being paid up to £400m to support England's test-and-trace service, and in doing so recruited a further 21 private sub-contractors to include Intellin, a company that employs around 500 staff. Intellin required that all employees utilise geo-tracking software due to allegations of some contact tracers working from abroad. Under GDPR, the rules are applied in controlling the flow of personal data and that Intellin are required to ensure adherence and compliance.

However, privacy campaigners question the ethics, legality and the manner of tracking staff and further highlight ongoing data protection concerns from the outset of the test-and-trace programme. If Intellin had to introduce tracking software to fulfil GDPR requirements, then it is feasible to expect that the other 20 companies may not have enabled this capability and should also introduce similar measures and geo-location data since the initial outbreak started (The Guardian, 2021).

In addition, ITV News reported that there was 'no clear evidence' that the UK Government's test-and-trace scheme reduced coronavirus infection levels. The Commons Public Accounts Committee (PAC) also questioned the effectiveness of the scheme stating that Test and Trace expenditure was justified and set up to prevent future lockdowns. However, England was still in its third lockdown (ITV, 2021). The British Prime Minister Boris Johnson defended the scheme by thanking the NHS and stating that the scheme enabled children to go back to school and enabled the cautious reopening of the economy in restarting our lives.

A main concern was that data failed to provide a measure of speed to the process from 'cough-to-contact', and therefore the public were unable to assess the programme's overall effectiveness. Further points included the scheme having difficulties with consistent supply and demand service delivery, resulting in sub-standard performance or surplus capacity. These types of 'failed' data are categorised such as 'Personal Data' and 'Special categories of personal data', which in turn are subject to the compliance and requirements of GDPR.

From a general data protection point of view, key information in addition to the normal circumstances is also being collected to include temperature, self-isolation confirmation and on-site visitors to a given premises. This information pertains to health status and is thus classified as both personal data and special categories of personal data (SCD). With the processing of SCD and personal data comes strict compliance to GDPR and its more local implementation framework, UK Data Protection Act 2018 (DPA) for example. However, to help organisations to understand the types, risks and mitigation steps of data processing, a Data Protection Impact Assessment (DPIA) is required if the data collected may result in disproportionate risk to an individual's rights and freedoms.

For example, in Bradford et al. (2020), an interesting article is presented that brings attention to how Digital Surveillance was a factor in containing several Covid-19 outbreaks in China, Israel, Singapore and South Korea. At the time, Google and Apple proposed building interfaces to enable Bluetooth contact tracing via iPhone and Android devices. Their paper tested the compatibility of proposed exposure notification systems with data protection regimes to include GDPR, US Health Insurance Portability and Accountability Act (HIPAA) and the California Consumer Privacy Act (CCPA). The authors found that GDPR's principle-based approach and blueprint enable compatibility with fundamental rights.

However, in contrast to GDPR, the more narrow US HIPAA and CCPA present gaps that could potentially expose weaknesses during time of uncertainty and emergency. The authors explain that although GDPR's scope is expansive, it is to be viewed as an advantage in conditions of uncertainty during a pandemic. GDPR has become the global gold standard since 2018, and especially in recent times, large social media companies have approached the Australian Government for example, asking for implementation of many GDPR elements when refreshing new privacy acts. Some GDPR alignments include changing 'personal information' to 'personal data' from within the Privacy Act, 1988, adopting 'multiple flexible legal bases for using or disclosing data' and implementing the right to information erasure (Facebook and Snap Inc, 2021).

Since government agencies and Silicon Valley (Big Tech Giants) collaboration raise privacy concerns, Lucivero et al. (2020) highlight the ethical challenges of mobile applications in utilising contact tracing for Covid-19. Questions are raised such as what defines a legitimate role for 'Big Tech' companies and what technology design safeguards are present for development and implementation. Attention is focused on what transparency, ethical oversight and data access [by whom] are and argues that if contact tracing is to be utilised, then it should be presented on a trial basis subject to independent evaluation and monitoring.

3   Research Challenges and Open Problems

To better understand the effect and performance of GDPR on BD in the context of Data Security and Data Privacy, we must first highlight and discuss recent and current directive interpretation and implementation research to include other legal frameworks with leading and emerging technologies such as AI, BC, Cloud Computing and Edge Computing. Reason being in that each technology compliments the other and is crucial to facilitating a large societal audience and platform to allow innovation, consistency, predictability and accountability to flourish on a global scale. The explosion of BD contributes to a wide range of industrial applications and domains to include the Health-Care sector.

3.1   Artificial Intelligence

Artificial Intelligence (AI) use cases include self-driving vehicles and drones, Cyber-Security (Botnet and Malware detection), Banking and Finance sectors.

For accurate and timely identification, diagnosis and remedies, AI in combination with BD supports the early prediction and treatment of a given diagnosis. For example, a standard AI method that classifies respiratory illnesses from BD is the Reverse Transcription Polymerase Chain Reaction (RT-PCR) (LabCorp COVID, 2020).

Due to the ongoing pandemic, more work has been undertaken to improve the RT-PCR method. However, due to cost, time and the instruments, materials and equipment needed, countries are limited in their RT-PCR testing capacities in quickly identifying and tracking illness in a global pandemic. Pandemics present immense challenges in the spreading of a disease (Covid-19), which requires a timely response from many different disciplines and fronts to include diagnostic modalities, drug development, resource allocation and planning.

In cases of quarantine implementation, this further complicates and stretches any efforts made, resulting in vast swathes of the population to be potentially furloughed in addition to being under evaluation for symptoms. In turn, this lack of personnel directly affects the access and testing abilities of health authorities which contribute data regarding the admittance of patient Accident and Emergency (A and E) and intensive care departments. In deploying AI as a tool to help mitigate the spread of disease, ethical and legal considerations are at the forefront of most people's minds when using this methodology (Williams et al., 2021).

An example in the Pharma Times reports on a British Court of Appeal hearing in regard to a controversial Immigration Exemption and data protection rights. The exemption is said to disapply particular data protection rights when processing immigration data. Healthcare providers, for example, may utilise this exemption in fulfilling Home Office requests in acquiring patient, immigration proceedings or investigations data.

However, due to the Covid-19 pandemic, this may dissuade immigrants from seeking healthcare support for fear of information being passed to the Home Office. The UK Government justification and defence of this exemption included the prevention of 'illegal immigrants learning they are to be deported and absconding'. If such a request is sent from the Home Office, then a balance of duty of care is undertaken with regard to a healthcare provider and their patients and reporting cases concerning UK citizens or immigrants. The Court of Appeal's decision will have significant ramifications for healthcare providers in receiving Home Office patient information requests (PharmaTimes, 2021).

3.2   Big Data

BD enables societies to produce extremely large and unprecedented amounts of data as a repercussion of data transmission architectures from portable and wearable sensors and devices, such as smartphones and sensors. Subject areas such as Data Cleaning highlight how the sheer velocity, volume, value and veracity of data present significant challenges in attaining and maintaining business advantage (Data Cleaning, 2020). Since the rapid adoption of the Internet of Things (IoT), BD in turn utilises advanced analytical tools and techniques to facilitate new knowledge and insights from the unstructured or structured data using Artificial Intelligence (AI) to identify patterns and trends within the data structures (Shastri, 2019).

Due to the pandemic, massive amounts of medical data continue to provide information on infection rates, which can be stored, trained and utilised to inform future preventative strategies. With this information, scientists, epidemiologists and health workers can make better informed decision in fighting the virus. Significant applications of BD include the capability of storing complete medical history records of all patients' travel history to help aid in early warning and risk assessment analysis (Haleem et al., 2020).

For example, BD coupled with AI acquires data from multiple formats which aid professionals in their decision-making abilities. Public data and electronic health records are analysed via intelligent analytics to support health services in the swift identification, diagnosis and remedies (i.e. drug and vaccine development) of a given illness including Covid-19 (IT Governance Privacy Team, 2017). As a crucial ability of BD, aggregated frameworks and complex simulation models are created to monitor, control and predict the spread of Covid-19 from data that is provided by various world aid and health agencies. This gives vital information to and from such agencies in declaring results and prescribing guidance to governments and their populace alongside undertaking better preventative measures (Barati et al., 2020).

However, it is extremely difficult to obtain a comprehensive and integrated view of (1) what personal data is using for storage from within an organisation, (2) making sure that an organisation understands the regulation content and (3) the production of the necessary records from data processing activities (Tran & Ngoc, 2020). Whilst GDPR compliance has enhanced the protection of personal data, i.e. Personal Identifiable Information (PII), sharing PII with add and marketing, collecting and sharing location, sharing PII of children, sharing with law enforcement and data aggregation, it remains a challenge as more work is necessary, particularly in the areas of granting users the right to edit, update and delete their data to entirely fulfil the GDPR promise (Zaeem & Barber, 2020).

Another challenge is linked user data that could link tuples/link data to groups of targeted users. A limitation here is that users themselves are not always aware of their own sensitive data. This warrants potential identification of attacks including confidentiality threats, highlighting further privacy preservation threats and the corresponding action for neutralising them, thus enhancing data profiling tools within self-service data preparation tools, especially for those targeted to end users and data officers (Caruccio et al., 2020).

3.3 Internet of Things

In an article from IoT Business News in March 21, 2021, Erik Fossum Færevaag, Founder and President of Disruptive Technologies, looks at how IoT has evolved during the Covid-19 pandemic. With the advent of Covid-19 coupled with the unprecedented remote working requirement, the need for smarter and faster technologies continues to push the IoT past all expectations to include managing a global crisis and adopting new work practices (Internet of Things, 2021). Cleaning staff, compliance monitoring, building management, supply chain and storage to include Smart City development are just a few of the ways businesses are looking into future development in the workplace and beyond.

In the UK, for example, the IoT provides a fundamental part in pandemic control via Track and Trace. In global vaccination delivery efforts in maintaining requisite vaccine temperature, cold chain monitoring is a system of mobile technologies used with IoT to monitor vaccine temperatures throughout the product journey to include, data loggers transmitting information during transport and storage. Via IoT, non-surgical robots are also being deployed to disinfect Covid hospital wards and mitigate contamination, using a UV light in destroying the virus. Other countries adapt smart monitoring to ensure cleaning is being appropriately undertaken, track how long and in what way. Businesses retaining core staff utilise in-house smart monitoring to help create a safer working environment (i.e. tracking use of space, hygiene management and personnel density) (Internet of Things, 2021).

3.4 Cloud Computing

With cloud computing comes the key benefit of cloud-based services for individual organisations in permitting fast implementation and up-scaling in various settings, as they do not require additional hardware or servers and can be implemented remotely. Huawei Technologies, for example, reported that a pneumonia diagnostic solution was deployed at a hospital in Ecuador in 14 hours; similarly, an application was deployed from Oklahoma State Department of Health for medical staff, designed to update people with reported symptoms of Covid-19 within 48 hours (Cresswell et al., 2021).

For example, global and digital inter-connectivity means in this context that data-pipelines, data-silos, data-warehouses and virtual data-centres monitor and facilitate business concepts such as Software as a Service (*SaaS:* Google Workspace, Dropbox, Salesforce, Cisco WebEx, Concur, GoToMeeting), Platform as a Service (*PaaS:* AWS Elastic Beanstalk, Windows Azure, Heroku, Force.com, Google App Engine, Apache Stratos, OpenShift) and Infrastructure as a Service (*IaaS:* DigitalOcean, Linode, Rackspace, Amazon Web Services (AWS), Cisco Metapod, Microsoft Azure, Google Compute Engine (GCE)) (Best CSP, 2021; SaaS vs PaaS vs IaaS, 2020).

A main benefit here is that the concepts *SaaS, PaaS and IaaS* are orchestrated by cloud service providers to include but not limited to companies like Microsoft Azure, Amazon Web Service (AWS), IBM Cloud, Google Cloud and Oracle Cloud Infrastructures. However, on premises solutions do permit the tailoring, piloting and contextualising of requirements for greater organisational control (Rawindran et al., 2021).

In using this type of rapid approach and implementation, although dealing with immediate challenges may have unintended consequences for patient safety, existing healthcare professionals and work practices, especially when implemented at large scales (Harrison et al., 2007). However, on premises solutions do permit the tailoring, piloting and contextualising of requirements for greater organisational control. This is an important factor, due to existing work alongside local health records identify a clear need to respond to a rapidly changing environment especially associated with Covid-19 (Reeves et al., 2020).

3.5   Edge Computing

In support of cloud computing, edge computing is a distributed framework to bring business and enterprise applications closer to their sources to include IoT devices and local edge servers (LESs) in an effort to increase response times, inform business intelligence and produce better overall bandwidth availability (IBM, 2020). The International Data Corporation (IDC), a global provider of market intelligence, advisory services and events for the information technology, telecommunications and consumer technology markets, estimates that there will be around an 800% growth in application numbers launched at the 'edge' by 2024. For example, as a result of Covid-19 and Brexit uncertainty, car production figures fell by around a third in 2020 to a level not seen in 36 years. Crucially, the ability to process and utilise data at a much faster rate and at a fraction of the price became an advantage, for example, in manufacturing and optimising product quality and yield that are of paramount importance. Edge technologies provide precision monitoring at the factory floor level on production lines (Techerati, 2021).

Shahaab (2020) states that trust in governments, corporations and the public services have been slowly receding over the last few decades due to lack of auditability and transparency. Also that since the advancement of consensus protocols (fundamental purpose and process of the Distributed Ledger Technology (DLT)), BC has evolved beyond notoriety in cryptocurrencies and is now contributing value across multiple sectors, platforms and advancements in Data Security and Privacy. The authors argue that consensus protocols should be appropriately and carefully applied per environment and sector requirements as no one protocol fits all (Shahaab et al., 2019).

For uniquely verifying, securing and sharing data, BC technologies are utilised in a multi-party, inter-organisational and cross-border fashion to carry out complex operational transactions. Global enterprises have invested in the technology to produce many proofs of concepts, and however real-world applications have been slow in its widespread use due to partner agreements regarding IP rights, governance and business models, with additional government impedance through regulation (Balasubramanian et al., 2021).

The Covid-19 pandemic has revealed weaknesses in our supply chains, efficient resource deployment and challenges in capturing and sharing data effectively to inform rapid decision-making. Consequently, the pandemic has also influenced the re-utilisation of BC solutions to help mitigate these supply chain weaknesses. Organisations such as the World Health Organization, Oracle, IBM and Microsoft are using BC solutions in partnership with government agencies, companies and other international health organisations in fighting the pandemic.

A recent example includes the BC-based open data hub, MiPasa. Its creators, HACERA, offer speed and precision in locating and detecting Covid-19 carriers and pandemic hot spots globally. MiPasa shares secured information between individuals, authorities and hospitals to aid in public health analysis. Additionally, in March 2020, the company Rapid Medical Parts was founded by Colonel James Allen Regenor, USAF (ret) whom previously developed a BC-powered platform to buy and sell traceable 3D printed parts, traditional parts and printing instructions; hence, using BC provided tamper-proof design and printing instructions. 12 days later, the US Pentagon awarded the company with a contract for converting sleep apnea machines into ventilators.

Due to the sudden demand for ventilators, Rapid Medical Parts provided additional parts at a tenth of the cost of purchasing a brand-new ventilator and create digital identifiers that cannot be traced to the source, thus preventing any tampering or the dissemination of personal data (HBR, 2021).

With regard to BC third parties, a challenge is in the conception of the mechanisms to register and authenticate the third parties that shall interact with the digital certificates system and how such a solution would integrate with the rest of the system preserving the privacy requirements (Molina et al., 2020).

3.6 5G Networks

As a rapid and effective means of curbing disease spread, social distancing measures meant that individuals, organisations and institutions had to depend on maintaining continuity via telecommunications to avoid complete operational shutdown. Fifth Generation Networks (5G), for example, endured immense connectivity pressures, and however the technology is still in its infancy to include enhanced Mobile Broad-Band (eMBB), massive machine-type communications (mMTCs) and Ultra-Reliable Low-Latency Communications (URLLCs), which together help mitigate pandemic associated challenges (Abubakar et al., 2020). In Gupta et al. (2020), the authors demonstrate a method of disease containment via BC and the utilisation of multi-swarm, Unmanned Aerial Vehicle (UAV). This method removes the physical human exposure element of the current pandemic. A UAV can also send vast quantities of data with a connection density of $107/km^2$ in real time via the 5G network (sub-6 GHz: mmWave-24.25 GHz and above) to ground-based cellular stations. In Gupta et al. (2020), the UAV concept presents fascinating potential and is quickly developing ahead in terms of product development due to innovation in the Covid-19 era.

To include other networks, 5G networks in themselves have the benefit of reliability, ultra-low-latency and with high bandwidth resolution. However, current UAV and Data Security and sharing methods highlighted by the authors compound efforts to carry out UAV deployment with a robust security strategy simultaneously. Therefore, a 6G (93 GHz–3 THz) network with intelligent connectivity and physical-level protocols with virtualisation of link (softwarized) are proposed via

BlockChain. Their results indicated that the proposed scheme performed better than 4G-/5G-based systems in terms of processing delay, throughput and packet loss reduction.

However, in terms of the GDPR, purpose limitations concerning UAVs should extend to health surveillance and interventions in a carefully controlled manner to ensure transparency and auditing capabilities as demonstrated by Barati and Rana (2020). To fully comply with GDPR and to maintain public trust, this concept may also appear intrusive by design.

Additionally, SaaS, PaaS and IaaS regulatory alignment are already said to be ahead of the GDPR framework function and purpose. This is not necessarily a good thing if influenced by external entities such as political rivalries and lobbyists. GDPR also impacts on data privacy and compliance for cloud service providers to include access to, storage, processing and transmission of personal data. GDPR can also be open to interpretation and ambiguous in implementing these operations, and so Barati proposes a transparent and automatic way of verification between CSPs and a user by utilising a BC-based virtual machine and a smart contract (Barati & Rana, 2020).

4 How Can We Use These New Applications During COVID-19 Whilst Being Complied to GDPR?

Here, the main critical component technologies are identified. The human element of utilising technology highlights working together in coopetition, the legal framework of which to operate by (GDPR) and the method of accountability upon a technological action or solution that can be trusted.

4.1 Coopetition

A main point here is promoting how to apply technologies in the first instance. In Elgazzar (2021), the author points out how coopetition can be used on the route to full cooperation and looks at how pharmaceutical and other interrelated enterprises interact during the pandemic using concepts such as the Nash Equilibrium, Quantum Game Theory, the Quantum Prisoner's Dilemma and concluded that in times such as the current pandemic, full cooperation or coopetition is necessary for sustainability and to control the spread of infection (Elgazzar, 2021). Crick (2020) illustrates how coopetition improves business performance and how the British retail sector adapted to changing market conditions with coopetition strategies whilst still being regulated under Covid-19 restrictions.

These ideas demonstrate and promote where organisational engagement in coopetition needs further research. Additional lessons may be learned to benefit a targeted approach to any future technology resource deployment strategy or service to encompass:

- **Crisis vs Continued Coopetition**: Coopetition performs well during a crisis to maximise resources. How can this be sustained in normal operating conditions for mutual benefit between companies, customers and stakeholders? Will companies continue to engage in coopetition, or will they pursue individualistic business models using their own resources and capabilities, post-crisis? (James & Crick, 2020).
- **New Customer-Centric Coopetition**: The creation of new areas where customers or citizens can co-opete, or organisations encourage internal coopetition. These challenges are how coopetition can be considered with customers at the centre (Walley, 2007).
- **Trust vs Distrust (long term vs short term)**: Understanding distrust which is often predicated upon previous encounters, particularly concerning category (military or non-military for neutrality), perceived ability and integrity (unreliable or risky), governance (lack of regulation and efficiency) or history at the individual or organisational level. Unfortunately, distrusted partners in coopetition are a key driver in encouraging organisations to implement strategies to manage uncertainty and to mitigate risk (Schiffling et al., 2020).

4.2 GDPR

Given the immense scale of the current pandemic mitigation strategies from around the globe, numerous countries look to adopt GDPR-like legal frameworks in their data protection governance. GDPR does have weaknesses such as unclear data storage specifications, yet it provides a gold standard to which should be treated as a tool to facilitate a flourishing culture. Additionally, in terms of data provenance, the current GDPR model does not present any information regarding its data sources and so could be perceived in opposition to the actual defined term of provenance which still present fundamental challenges.

GDPR, for example, claims to be the 'toughest privacy and security law in the world', GDPR (2018), which imposes and obligates organisations in executing data privacy and security measures for its citizens living within the EU. However, current models of GDPR data provenance are stated to not be in full compliance and so set out a number of changes to existing research models to include high-level granularity oversight with meta-data, undertaking inter-controller audits and highlighting the lack of appropriate safeguarding in regard to the data controller and their availability in legally transferring personal data.

4.3   Blockchain (BC)

In terms of transparency and enforcement, a case study by Barati (2020) demonstrates the encoding of GDPR rules for each operation carried out. The data is recorded into a BC for auditing purposes and to demonstrate how these rules can appear as OPCODES in smart contracts in verifying provider operations (read, write, execution and transfer) on user data.

A formal model is proposed for the automation of operational verification in smart devices via three smart contracts. They conclude that there is a direct relationship between the number of GDPR violations detected and the fee paid by the user. However, GDPR protocols conflict with BC protocols, thus presenting enormous challenges to include how GDPR rules, particularly those legislated for cloud technologies, are realised in smart contracts and how the GDPR rule 'the right to be forgotten' cannot be fully supported, since the rule currently has a conflict with the immutability of a BC network (Barati & Rana, 2020).

In Campagna et al. (2020), the authors present a paper focusing on enforcing compliance whilst contributing to a research branch promoting *data provenance* in achieving transparency objectives set out in GDPR. In an article by Barati et al. (2020), the authors test the performance cost and scalability of a BC peer-to-peer network assessing the visibility of how smart devices use personal data. This highlights how BC is immutable and its utilisation for auditing purposes in the provision of privacy and transparency.

In addition, a focus is that smart contracts are already deployed across IoT networks via smart devices hosted by cloud services, stating the benefits of GDPR to include the provision of consent in performing data manipulation between associated parties via the cloud and BC.

4.4   Cloud Computing

Technological development alongside legal frameworks often prove challenging. However, additional requirements are necessary for regulating the consistency, security and quality of information dissemination on a global scale. This is typically dependent upon agreed national and international legal frameworks and directives, to carry out maintenance and intervention on behalf of the public. With the utility of the Cloud, BD and AI in a global pandemic, for example, additional risks continue to evolve in the effectiveness and delivery of the most critical component to any plan, real-time accurate information.

For example, in Pham et al. (2019), the authors propose an optimisation computation offloading framework, which allocates resources via multiple Low Energy Servers (LESs) to facilitate mobile-edge computing. A main objective here is to minimise the overall system computational overheads by focusing on computational decision-making, user power transmission and server actualisation cost, of which is originally an in-combination problem. The idea was to focus on individual components such as joint resource allocation and computation offloading.

In Barati and Rana (2020), the authors bring attention to the utilisation of a BC-based virtual machine for auditing purposes. Barati (2020) presents an abstract model and case study which demonstrates how to deploy smart contracts for cloud providers and how to utilise in practice. The model uses collections of design patterns and smart contracts for provider verification to include the read, write, execution and transfer concerning user data.

For this framework to be fully realised and exploited globally, partners in industry and organisations should facilitate aligned coopetitive strategies to enable accuracy, fairness, lawfulness and in terms of Cyber-Security, the *Confidentiality* and *Integrity* and *Availability* of such services. It is of critical importance that interactions with institutions, social behaviours, interpersonal relationships, technology and other domains can allow for innovation on resource, trust and information strategies concerning humanitarian operations (HOs) (Walley, 2007).

5   Conclusion

Early on in this chapter, we carefully defined and justified that in an effort to demonstrate GDPR principles and personal data protection outcomes in an era of Covid-19, numerous projects symbiotic to BD technologies were explored across a diverse range of source materials to bring attention to new ideas and concepts.

From the evidence gathered, a main pandemic operational issue for GDPR and BD in the promotion of operational trust, technology development speed and the resistance of legal and procedural precedence is the terms of delivering upon humanitarian operations and logistics in emergency situations. Trust-less mutual agreements such as smart contracts offer traceability on a level that has not been seen before. As shown in various academic works, a given problem only has to be catalogued in a way that enables transparency.

In facilitating concepts like BC, BD and GDPR working together in an era of Covid-19 to benefit all, clearly, levels of implicit trust are a necessary factor in the successful deployment of any coopetitive strategy. It is clear that GDPR is a step in the right direction with regard to data protection. Yet, a factor that has been shown again and again in the media is that at times of emergencies and large-scale disasters, that organisations external to the EU find it hard to align or simply comply with the framework due to geographical location, resident country IT infrastructure investment or political strategies of misinformation seeking opportunism. Scalability still presents significant challenges due to a lack of will.

It is clear that to implement a successful coopetative strategy, rapid and effective communication between business, organisations and the general public of the affected areas is key. A main arm of delivering a robust and flexible strategy is by augmenting technology more effectively in operations such as planning, resource deployment and the overall monitoring and project maintenance tasks.

6 Recommendations

For GDPR to accommodate Information Technologies successfully, there would need to be a significant development in legal frameworks research and development motivations. The creation of new areas where customers or citizens can co-opete and encourage organisations internal coopetition may be the key, as coopetition performs well during a crisis. Therefore, a more holistic understanding and approach is needed, to understand collective erosion of trust and its inter-relating effects on such domains and their citizens (public), as trust is something identified as a vital means to communicate social and health behaviour, especially during a global pandemic.

Additional and external BC operations provide excellent opportunities for auditing and compliance (OPCODES), and thus setting up a smart contract may ensure a more effective approach to GDPR compliance. This could be easily implemented to support BD and GDPR accountability and transparency in carrying out justice. Covid-19 has highlighted how important the balance is between privacy and defending against future disasters and pandemics. GDPR would need to be continually evolving in facilitating these challenges.

References

Abubakar, A. I., Omeke, K. G., Ozturk, M., Hussain, S., & Imran, M. A. (2020). The role of artificial intelligence driven 5G networks in COVID-19 outbreak: Opportunities, challenges, and future outlook. *Frontiers in Communications and Networks, 1*, 4. https://www.frontiersin.org/articles/10.3389/frcmn.2020.575065/full. Accessed March 09, 2021.

Balasubramanian, R., Prakash, E., Khan, I., & Platts, J. (2021). Blockchain technology for healthcare. In *AMI 2021 - The 5th Advances in Management and Innovation Conference 2021.* Cardiff Metropolitan University.

Barati, M., & Rana, O. (2020). Tracking GDPR compliance in cloud-based service delivery. *IEEE Transactions on Services Computing*, 1–1. https://doi.org/10.1109/TSC.2020.2999559

Barati, M., Rana, O., Petri, I., & Theodorakopoulos, G. (2020). GDPR compliance verification in internet of things. *IEEE Access, 8*, 119697–119709.

Best Cloud Computing Services of 2021: For Digital Transformation. (2021). https://www.techradar.com/uk/best/best-cloud-computing-services. Accessed November 27, 2020.

Bradford, L., Aboy, M., & Liddell, K. (2020). Covid-19 contact tracing apps: A stress test for privacy, the GDPR, and data protection regimes. *Journal of Law and the Biosciences, 7*(1), lsaa034.

Campagna, D., da Silva, P., Altigran, S., & Braganholo, V. (2020). Achieving GDPR compliance through provenance: An extended model.

Caruccio, L., Desiato, D., Polese, G., & Tortora, G. (2020). GDPR compliant information confidentiality preservation in big data processing. *IEEE Access, 8*, 205034–205050.

Covid: 'No clear evidence' Test and Trace scheme works, MPs say in critical report. (2021). https://www.itv.com/news/2021-03-10/mps-say-no-clear-evidence-test-and-trace-scheme-works-in-critical-report. Accessed March 12, 2021.

Covid-19 Alert. (2020). https://www.fitfortravel.nhs.uk/advice/disease-prevention-advice/coronavirus-disease-covid-19. Accessed January 21, 2021.

Cresswell, K., Williams, R., & Sheikh, A. (2021). Using cloud technology in health care during the covid-19 pandemic. *The Lancet Digital Health, 3*(1), e4–e5.

Data Cleaning: Challenges and Novel Solutions. (2020). https://doi.org/10.25401/cardiffmet. 12252698.v1. Accessed December 21, 2020.

Edge Computing After Covid-19. (2021). https://www.techerati.com/features-hub/opinions/edge-computing-after-covid-19/. Accessed April 12, 2021.

Elgazzar, A. S. (2021). Coopetition in quantum prisoner's dilemma and covid-19. *Quantum Information Processing, 20*(3), 1–13.

Facebook and Snap Inc call for a GDPR-aligned Australian Privacy Act. (2021). https://www.zdnet.com/article/facebook-and-snap-inc-call-for-a-gdpr-aligned-australian-privacy-act/. Accessed February 12, 2021.

Goddard, M. (2017). The EU General Data Protection Regulation (GDPR): European regulation that has a global impact. *International Journal of Market Research, 59*(6), 703–705.

Gupta, R., Kumari, A., Tanwar, S., & Kumar, N. (2020). Blockchain-envisioned softwarized multi-swarming UAVs to tackle COVID-I9 situations. *IEEE Network, 35*(2), 160–167.

Haleem, A., Javaid, M., Khan, H. I., & Vaishya, R. (2020). Significant applications of big data in covid-19 pandemic. *Indian Journal of Orthopaedics, 54*, 526–528.

Harrison, M. I., Koppel, R., & Bar-Lev, S. (2007). Unintended consequences of information technologies in health care—an interactive sociotechnical analysis. *Journal of the American Medical Informatics Association, 14*(5), 542–549.

How Covid-19 accelerated the dominance of the Internet of Things. (2021). https://iotbusinessnews.com/2021/03/21/52141-how-covid-19-accelerated-the-dominance-of-the-internet-of-things/, 2021. Accessed April 12, 2021.

How the Pandemic Is Pushing Blockchain Forward. (2021). https://hbr.org/2020/04/how-the-pandemic-is-pushing-blockchain-forward. Accessed February 12, 2021.

IT Governance Privacy Team. (2017). EU: General data protection regulation-an implementation and compliance guidance.

James, M., & Crick, D. (2020). Coopetition and COVID-19: Collaborative business-to-business marketing strategies in a pandemic crisis. *Industrial Marketing Management, 88*, 206–213.

LabCorp COVID. (2020). RT-PCR test EUA Summary. *Accelerated Emergency Use Authorization (EUA) Summary COVID-19 RT-PCR Test (Laboratory Corporation of America)*. https://www.fda.gov/media/136151/download. Accessed November 27, 2020.

Living with COVID-19: Opportunism and International Security. (2020). https://www.globalhealthhub.de/en/event/living-covid-19-opportunism-and-international-security. Accessed October 14, 2020.

Lucivero, F., Hallowell, N., Johnson, S., Prainsack, B., Samuel, G., & Sharon, T. (2020). Covid-19 and contact tracing apps: Ethical challenges for a social experiment on a global scale. *Journal of Bioethical Inquiry, 17*(4), 835–839.

Medrano, N., & Olarte-Pascual, C. (2016). The effects on the crisis on marketing innovation: An application for Spain. *Journal of Business and Industrial Marketing*.

Molina, F., Betarte, G., & Luna, C. (2020). A Blockchain based and GDPR-compliant design of a system for digital education certificates. Preprint, arXiv:2010.12980.

Pham, Q., Fang, F., Ha V. N., Piran, J., Le, M., Le, L. B., Hwang, W.-J., & Ding, Z. (2019). A survey of multi-access edge computing in 5G and beyond: Fundamentals, technology integration, and state-of-the-art. Preprint, arXiv:1906.08452.

Poplavska, E., Norton, T., Wilson, S., & Sadeh, N. (2020). From prescription to description: Mapping the GDPR to a privacy policy corpus annotation scheme. In *Legal Knowledge and Information Systems* (pp. 243–246). IOS Press.

Rawindran, N., Jayal, A., & Prakash, E. (2021). Artificial intelligence and machine learning within the context of cyber security used in the UK SME sector. In *AMI 2021 - The 5th Advances in Management and Innovation Conference 2021*. Cardiff Metropolitan University.

Reeves, J. J., Hollandsworth, H. M., Torriani, F. J., Taplitz, R., Abeles, S., Tai-Seale, M., Millen, M., Clay, J. B., & Longhurst, A. C. (2020). Rapid response to covid-19: Health informatics support for outbreak management in an academic health system. *Journal of the American Medical Informatics Association, 27*(6), 853–859.

Revealed: Some of England's Covid contact tracers working from abroad. (2021). https://www.theguardian.com/world/2021/feb/09/revealed-uk-covid-contact-tracers-working-from-abroad. Accessed February 18, 2021.

SaaS vs PaaS vs IaaS: What's the difference and how to choose. (2020). https://www.bmc.com/blogs/saas-vs-paas-vs-iaas-whats-the-difference-and-how-to-choose/#ref1. Accessed December 27, 2020.

Schiffling, S., Hannibal, C., Fan, Y., & Tickle, M. (2020, June). Coopetition in temporary contexts: Examining swift trust and swift distrust in humanitarian operations. *International Journal of Operations and Production Management.*

Shahaab, A., Lidgey, B., Hewage, C., & Khan, I. (2019). Applicability and appropriateness of distributed ledgers consensus protocols in public and private sectors: A systematic review. *IEEE Access, 7,* 43622–43636.

Shahaab, A., Maude, R., Hewage, C., & Khan, I. (2020). Blockchain-A panacea for trust challenges in public services? A socio-technical perspective. *The Journal of The British Blockchain Association,* 14128. https://doi.org/10.31585/jbba-3-2-(6)2020

Shastri, S., Banakar, V., Wasserman, M., Kumar, A., & Chidambaram, V. (2019). Understanding and benchmarking the impact of GDPR on database systems. Preprint, arXiv:1910.00728.

The Immigration Exemption and the GDPR. (2021). http://www.pharmatimes.com/web_exclusives/The_Immigration_Exemption_and_the_GDPR_1364475, 2021. Accessed March 02, 2021.

Thorgren, F. (2019). Blockkedjeteknik och dataskyddsför-ordningens krav på radering.

Tran, J., & Ngoc, C. (2020). *GDPR handbook for record of processing activities. Case: The color club A/S.*

Walley, K. (2007, July). Coopetition: An introduction to the subject and an agenda for research. *International Studies of Management and Organization, 37,* 11–31.

What is edge computing? (2020). https://www.ibm.com/uk-en/cloud/what-is-edge-computing. Accessed December 21, 2020.

What is GDPR, the EU's new data protection law? (2018). https://gdpr.eu/what-is-gdpr/. Accessed October 22, 2020.

WhatsApp policy, Indians apprehensive about privacy, says CJI. (2021). https://www.thehindu.com/news/national/sc-notice-to-centre-whatsapp-on-plea-alleging-lower-standards-of-privacy-for-indian-users/article33840452.ece?homepage=true. Accessed February 15, 2021.

WHO Coronavirus Disease (Covid-19) Dashboard. (2021). https://covid19.who.int/. Accessed October 22, 2020.

Williams, C. M., Chaturvedi, R., Urman, R. D., Waterman, R. S., & Gabriel, R. A. (2021). Artificial intelligence and a pandemic: An analysis of the potential uses and drawbacks. *Journal of Medical Systems, 45*(3), 26.

Wylde, V., Prakash, E., Chaminda, H., & Platts, J. (2021). Covid-19 crisis: Is our personal data likely to be breached? In *AMI 2021 - The 5th Advances in Management and Innovation Conference 2021.* Cardiff Metropolitan University.

Zaeem, R. N., & Barber, S. K. (2020). The effect of the GDPR on privacy policies: Recent progress and future promise. *ACM Transactions on Management Information Systems (TMIS), 12*(1), 1–20.

Privacy and Security Challenges and Opportunities for IoT Technologies During and Beyond COVID-19

V. Bentotahewa, M. Yousif, C. Hewage, L. Nawaf, and J. Williams

1 Introduction

The Internet of Things (IoT) technology sees extensive growth with the increased number of smart devices connected via the Internet. The global market for IoT solutions is expected to grow to around 1.6 trillion USD by 2025 (Statista, 2021). These predicted trends will give rise to the expansion of opportunities created by the COVID-19 pandemic. The IoT solutions such as remote health monitoring and contact tracing enabled authorities to successfully manage the spread of the coronavirus. However, wider deployment of IoT-inspired technologies faces challenging obstacles such as privacy and security concerns. This chapter uptakes a comprehensive review of these challenges and an in-depth analysis of the issue.

1.1 IoT Role During COVID-19 Pandemic

Coronavirus disease 2019 (COVID-19) is an infectious illness caused by a novel and newly discovered coronavirus (W. H. Organization, 2021; COVID-19, 2021). Some symptoms of the disease are shortness of breath, chest pain, and fever. On the 11th of March 2020, the World Health Organization (WHO) declared a global pandemic

The original version of this chapter was revised: Second author's name has been corrected. The correction to this chapter is available at https://doi.org/10.1007/978-3-030-91218-5_11

V. Bentotahewa (✉) · M. Yousif · C. Hewage · L. Nawaf · J. Williams
School of Technologies, Cardiff Metropolitan University, Cardiff, UK
e-mail: vibentotahewa@cardiffmet.ac.uk

due to the COVID-19 outbreak. To reduce social contact during the pandemic, 'some businesses must remain closed or follow restrictions on how they provide goods and services' ((COVID-19) Coronavirus restrictions: What you can and cannot do—GOV.UK, 2021).

Harvard Medical School has highlighted that in certain COVID-19-related heart injury patients, the initial symptoms might have occurred in several forms (Pesheva, 2020). Those without previous underlying cardiac problems might remain healthy, while in others, oxygen supply failure to heart muscles might cause heart damage. In the context of COVID-19, the contributory factor would be the imbalance between 'supply and demand' for oxygen to the heart. In the process of stabilising oxygen levels in the body, IoT played a key role in efficiently managing the pulse oximeter, nebulisers and oxygen tanks.

IoT technology has been used extensively for many purposes across diverse sectors during the pandemic as was referred to earlier, and their applications and frameworks have enabled successful management of the pandemic. Prior to the onset of the COVID-19 pandemic, IoT had been linked to certain key areas or catch phrases such as smart homes, self-driving cars, smart metering, etc. However, in the aftermath of the pandemic, IoT was put into effective use across a wide range of sectors for purposes such as contact tracing, retail and hospitality. The key IoT sectors affected by the pandemic, the economic/social impact and technology readiness levels (TRL) are discussed in (Yousif et al., 2021). An elaboration of the industries affected and the IoT solutions used during the pandemic are set out in the following subsections.

Affected Industries

Different industries, such as the hospitality sector and the restaurant industry, were affected by the COVID-19 pandemic. The knock-on effect was felt in small and medium enterprise sectors, and consequently, they were badly hit. For instance, 3% of restaurants remained permanently closed (Song et al., 2021); tourism sector (Altuntas & Gok, 2021) because of travel restrictions and freedom of movement due to social distancing rules; airline industry because of operational changes in air travel and airports and the travel agencies for the same reasons; agricultural sector (Dutta & Mitra, 2021), on which other sectors, mentioned above, depended on; the retail industry on which the consumers relied on to sustain their livelihood (COVID-19 and the retail sector: Impact and policy responses, 2020); education delivery system switching to virtual distance learning (Ilieva & Yankova, 2020); and healthcare services overwhelmed by the virus and COVID-19 cases. These are some of the examples of the affected industries, and the IoT solutions applied are highlighted in each case. The selected industries were chosen based on the background knowledge of the authors and the reviewed articles (Panchal, 2019; Self Checkout Systems in 2021: Comprehensive Guide, 2021; Triax Technologies, 2021; Covid-19 Temperature Screening Service & Test | Metro Security, 2021; Obaidat et al., 2020; Waheed et al., 2020; Berkay Celik et al., 2019).

IoT Solutions

In the agriculture industry and with the use of sensors, IoT-based smart farms could survive. IoT smart farms allow data collection, tracking remote monitoring and remote control. The use of IoT in agriculture makes factories more efficient, optimises treatment and input required and efficient water use and will make the environment better (Dutta & Mitra, 2021). By implementing IoT technologies such as drones and sensors, we can monitor crop health, seed inspection, seed harvesting and soil examination. The author in (Rowan & Galanakis, 2020) proposes the use of immersive technologies and information and communications technology (ICT) for remote end-user applications, also, to inform disruptive innovations.

The author in (Pillai et al., 2021) describes how IoT devices can lead to a hassle-free post-checkout sanitisation that eliminates human-to-human interaction and enabled service reconfiguration, based on customer preference survey of consumer behaviour and predictions in the hospitality industry. In addition, improvements to workplace safety can be made by installing real-time alarms to alert emergencies. IoT can also be used to ensure maintenance of hygiene standards in the sales outlets (cleanliness of restaurant tables, sanitiser solution concentration, contactless payment and communication) and adherence to social distancing rules (Suleman, 2021; Embree, 2021) to minimise the need for manual interventions. The use of IoT retail self-checkouts such as kiosks, IoT-automated systems such as Amazon warehouse and RFID inventory tracking can help limit interaction between humans, thereby avoiding human error and excess staff numbers, and enhance supply chain management with inventory, delivery and storage (Panchal, 2019; Self Checkout Systems in 2021: Comprehensive Guide, 2021). There are diverse types of IoT wearables and devices for contact tracing and temperature screening used in the healthcare industry to ensure social distancing, accurate diagnosis, tracking and health monitoring and provide exposure notification (Triax Technologies, 2021; Covid-19 Temperature Screening Service & Test | Metro Security, 2021).

Figure 1 provides a summary of IoT solutions that are used in different industries.

This review provides an in-depth understanding of the main IoT sectors that played a vital role in managing the global pandemic and their potential applications in the post-COVID-19 future. Authors expect the usage of IoT-based technologies and applications to increase significantly during the next normal matching lifestyle patterns such as working from home, distance learning and telemedicine that have emerged during the pandemic.

The potential for this technology is immense, but the challenges are likely to be equally immense. Amongst other concerning issues, energy requirements of these IoT devices and privacy and security are key priorities for consideration (Obaidat et al., 2020; Waheed et al., 2020; Berkay Celik et al., 2019). However, there is a lack of COVID-19 pandemic-relevant literature published on the issues touched upon earlier. Therefore, in the absence of informative literature, the primary focus of this chapter will be on privacy and security issues associated with IoT data collection (Big Data) and security challenges. Section 1.2 summarises the privacy and security challenges of IoT.

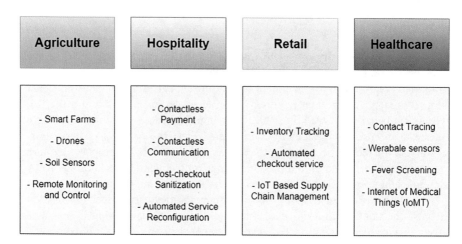

Fig. 1 Key IoT solutions during the COVID-19 pandemic

1.2 Privacy and Security Concerns of IoT

Since the Internet of Things (IoT) came to being, its applications and the range of connected devices have multiplied, and in parallel, the expanding usage of IoT also induces many technical challenges potentially threatening the security and privacy of IoT end-users. Therefore, there is an imperative requirement to put in place risk mitigation solutions, sooner than later.

In the IoT environment, while safeguarding online security remains a major concern and a challenge, preserving privacy will also remain a significant challenge needing added attention. As an example, the privacy of the IoT end-users could be at risk if personal data happens to be leaked to unauthorised persons or even through a security breach in the IoT (devices). Such incidents would potentially allow the attacker access to IoT end-user data without being tracked or traced by (face recognition) security cameras located in smart homes. Given the heterogeneity of IoT-connected devices and in-built vulnerabilities of hardware and software in some of them, safeguarding end-user privacy might face many security challenges (Bertino, 2016).

There are reported studies focusing on the privacy and security challenges of IoT (Obaidat et al., 2020; Waheed et al., 2020; Berkay Celik et al., 2019). However, this chapter provides an in-depth analysis of these important challenges especially in the aftermath of COVID-19. The discussion in Sect. 2 of this chapter focuses on the large volume of information generated through IoT devices, the analysis of security and privacy challenges associated with Big Data and the provision of legal and policy solutions to protect privacy for maintaining trust between the data subject and data controller. IoT threats, security challenges and proposed solutions are discussed in Sect. 3 of this chapter. In addition, the impact of COVID-19 and the role of IoT in different industries are highlighted at the end of Sect. 3.

1.3　Study Methodology

The aim of this study is to review the privacy and security challenges of IoT technologies for the next normal. The research objectives of this study are listed below:

(a) Identification and in-depth analysis of privacy and data protection challenges associated with Big Data generated via IoT technologies.
(b) Analysis and discussion of security challenges for IoT technologies for the next normal and finally the identification and analysis of the best practices and code of practices for IoT technologies.

The review was carried out by using publicly available secondary data sources that explore and discuss different aspects of IoT technologies in diverse sectors. The main data sources used in this review are the Scopus library, Web of Science citation database, ACM library, IEEE Xplore, Google Scholar and ResearchGate. A number of keyword searches were used to find relevant studies and reviews necessary to answer the research questions of this study. An exclusion criterion was not used to provide a wider overview of the issue. In addition to the initial research by the authors, recommendations by previously published research, tutorials, surveys and reviews were used to select the prominent privacy and security challenges to focus on in this study.

1.4　Structure of the Chapter

This chapter is organised as follows: Sect. 2 discusses IoT vs. data protection. Security architectures are discussed in Sect. 3. Section 4 summarises the future privacy and security landscape for IoT. The conclusion of the study is provided in Sect. 5.

2　Data Protection vs. IoT

IoT technology has been used widely during the COVID-19 pandemic for the purpose of mitigating and preventing the spread of the coronavirus. These Internet-connected devices did serve the purpose, but they also gave rise to an upsurge of privacy and security risks associated with the collection of a large volume of data. Section 2 is dedicated to investigating IoT-generated Big Data and what actions could be taken to protect them. Section 2.1 focuses on literature-based definitions for Big Data generated by IoT, associated threats and the importance of protecting Big Data. The authors have dedicated Sect. 2.2 to highlight the data protection challenges and existing solutions to overcome potential challenges. In Sect. 2.3, the authors have flagged up relevant data laws associated with Big Data in parallel with GDPR. Section 2.4 highlights policy mechanisms and their purpose in the context of Big Data.

2.1 Usefulness and Security of Big Data Generated by IoT

The question that is often asked by those who are not familiar with modern tech jargon is 'what is Big Data'. To explain it in simple terms, it is a vast amount of information collected for understanding and decision-making purposes using innovative forms of information processing (Wu et al., 2014). In professional literature, the definition of Big Data refers to the volume of data collected, the variety of sources and the speed of analysis and interpretation that could be achieved through the analytical process (Erevelles et al., 2016). Data collected in this way have the capacity to reveal information about individuals in terms of their habits, location, interests and a host of other personal information and varying preferences that are stored in the systems for usage with ease. While there is no single definition of Big Data, the Information Commissioner's Office (ICO) believes that it is useful to regard Big Data as data which, due to several varying characteristics, is difficult to analyse using traditional data analysis methods (Richard, n.d.).

Big Data comes in various formats (Fig. 2), such as cell phone location information, CCTV recordings, social media contents from a variety of sources and satellite images (Oussous et al., 2018), and handling them is a significant challenge. Primarily, data that relates to an identifiable living individual is considered as Big Data in (Article 4(1), General Data Protection Regulation (GDPR) (Intersoft Consulting, 2019a), but not all Big Data, for example, climate and weather data, is

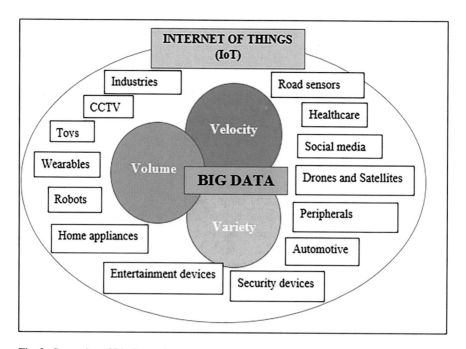

Fig. 2 Generation of Big Data. (Source: Authors, 2021)

not personal data (Richard, n.d.). Reports highlight the significant increase in the frequency of data breaches since 2015 (60% in the USA only) (Tawalbeh et al., 2020). In 2016, the world was introduced to the security risks and vulnerabilities associated with smart technology aftermath of the Mirai IoT botnet Denial-of-Service (DoS) attack which caused widespread Internet outages throughout the USA and Europe (Rosenthal & Oberly, 2020). A survey conducted in Japan, Canada, the UK, Australia, the USA and France has revealed that 63% of the IoT consumers felt these devices could not be trusted due to security inadequaties (Tawalbeh et al., 2020). Also, research findings have highlighted that 90% of consumers did not seem to have confidence in IoT cybersecurity (Tawalbeh et al., 2020).

There is no doubt that connected things in various sectors do bring tangible benefits that make life better, but also, they carry with them serious concerns about data security. There is no single magical solution to solve the identified Big Data security and privacy challenges. There are various challenges connected with data collecting, processing and storing. A vast volume of data become irrelevant unless they are processed to get something useful out of them. Therefore, it is important to ensure that the sensors function properly and the quality of the data coming for analysis is reliable and not spoiled by factors such as environmental conditions and sensor malfunction/breakdown.

Security of Big Data and privacy is an essential element that will ensure data trustworthiness in the data collection process and usage. In general, the majority of data breaches and IoT attacks happen due to a lack of user awareness (Jurcut et al., 2020; Tawalbeh et al., 2020). Therefore, documented user guidelines should be compulsory to strengthen security awareness. It has been reported that IoT security measures and guidelines had not been usually mentioned when the users purchased these devices (Tawalbeh et al., 2020). To avoid any controversies, the device manufacturers ought to take the lead to bring potential IoT threats to the attention of the user, and the organisations should produce a package of effective training programs to enhance security awareness. In a positive move, in contrast, data protection authorities point out that, like any other form of data processing, Big Data falls within the framework of data protection law and must comply with data protection legislation in accordance with GDPR which was established with technological advancements in mind.

2.2 Big Data Protection Challenges

In practice, data protection and security become extremely challenging in an IoT environment, as a communication interface between objects and persons is at the core of the system, without human intervention. Given the pace of change, it is not surprising that there is little evidence to presume that data protection is keeping up with the pace of change. Even though when legislative drafters demonstrate their awareness of specific concerns in processing data on a large scale, their understanding of risk implications may not be sufficient in practice.

Big Data applications typically tend to collect data from diverse sources, without careful verification of the relevance or accuracy of the data thus collected (Mortier et al., 2021). Google's unsuccessful attempts at health diagnostic and, most recently, the use of analytics to predict the US election results (McDermott, 2017) can be taken as good examples of the inaccuracy of Big Data. On that basis, the accuracy principle can be challenged as the GDPR underscores the importance of accuracy (Intersoft Consulting, 2019b) in personal data.

The GDPR applies to the processing of personal data, regardless of whether the processing takes place in the EU (Intersoft Consulting, 2019c). The controllers and processors and those acting as controllers of Big Data as well as those acting as processors on their behalf are obliged to comply with GDPR. The application of data protection principles could be challenging when using personal data in the Big Data context, especially where it involves the use of techniques made possible by AI. These implications arise not only from the volume of data but also from the ways in which data is generated, stored and processed.

The creation of personal data in a vast amount through Big Data techniques allows organisations to combine different datasets, and that is likely to increase the capability of data to identify living individuals in new ways (Brogan, 2019). As a result, the capacity to mine and analyse datasets increases in volumes, variability and velocity effectively giving rise to an exponentially increased volume of personal data. To overcome the challenges, in the context of Big Data, it is advisable to consider whether personal data can be fully anonymised. The GDPR specifies that the principles of data protection should not apply to anonymous information that does not relate to an identified or identifiable real person or personal data classified 'anonymous' (Intersoft Consulting, 2019d) in such a manner data subject's information is not protected under the GDPR. Therefore, organisations who use anonymised data are expected to verify that they had carried out a robust assessment of the risk of re-identification and adopted proportionate solutions (Information Commissioner's Office, n.d.-a). This may involve a range of technical measures, such as data masking, pseudonymisation, aggregation and banding, as well as legal and organisational safeguards (Information Commissioner's Office, n.d.-a).

The UK Anonymisation Network (UKAN) plays a significant role in providing expert advice on anonymisation techniques (UK Anonymization Network, 2021). It also enables the organisation to reassure people that collected data capable of identifying them will not be used for Big Data analytics (Information Commissioner's Office, n.d.-a). This is an important criterion for building trust and in taking Big Data forward. However, some commentators have made references to examples where it had been possible to identify individuals in anonymised datasets but had concluded that anonymisation was becoming increasingly ineffective (Information Commissioner's Office, n.d.-a, n.d.-b). However, personal data that had been pseudonymised, in other words, identify an individual in conjunction with additional information could still be possible and will remain as classed personal data (Richard, n.d.).

In the ICO Big Data Paper 2017, the ICO emphasises the importance of fairness, transparency and the need for meeting the data subject's reasonable expectations in Big Data processing (Information Commissioner's Office, n.d.-b). However, as a vast amount of data are processed through massive networks in a daily basis, there is limited transparency in how these algorithms work and how data is processed. Furthermore, the ICO Big Data Paper 2017 notes that the complexity of Big Data analytics can lead to mistrust and potentially be a barrier to data sharing, particularly both in the public and the private sectors. This can lead to reduced competitiveness as a negative perception of the consumer will impact trustworthiness (Richard, n.d.). Therefore, in the Big Data context, privacy notices (Information Commissioner's Office, 2020) serve as an important means of providing transparency, while also the consent factor (Intersoft Consulting, 2019e) has been the most reliable in ensuring transparency. The ICO Big Data Paper 2017 makes it clear that the complexity of Big Data should not be taken as an excuse for failing to obtain consent if and when required to do so (Information Commissioner's Office, n.d.-b). The GDPR also follows this approach by asserting that data processing is conditional on obtaining prior consent from the data subject (Intersoft Consulting, 2019e). However, the assertion to obtain consent for processing might not turn out to be a workable solution in all circumstances because of the complexity of the analytics. A study in the USA suggests that companies overestimate customers' concerns about the use of their personal data. It claims that in reality, people are primarily concerned about what the organisations plan to do with their data (Information Commissioner's Office, n.d.-a). This leads to the point that personal security remains uppermost in their thinking. Then it is arguably clear that emphasis should be on the data collection process and use rather than focusing on controlling what happens after data is collected. Therefore, where an organisation is relying on consent in the Big Data context, people must have an understanding of how the organisation will use their data and a clear indication of consent given for the intended purpose only. To determine the intended purpose compatibility of data originally collected and used will increasingly become challenging with Big Data. If an organisation had collected personal data for one purpose and then decided to start analysing for completely different purposes, the users need to be made aware of the changes, and, where necessary, further consent needs to be obtained.

Connected things generate terabytes of data; therefore, deciding which data to store and which to drop is a demanding task in data minimisation. The custodians of stored data may need to retain them for use over a long period for use in the future. The challenge is to secure critical data from criminals and unauthorised access. Any breaches will compromise the privacy of the users and have a negative impact on the image of the custodian, affecting trustworthiness, and the users will lose faith not only in the organisation but also in the system. According to an assumption that emerged in 2006, there were notable concerns about invasion of privacy amongst the adult population than the younger generation who felt comfortable about revealing their personal information (Maple, 2017). But there had been proposed changes, and the Oxford Internet Institute had released a report, in which it had stated that young people were found to be more likely to take action to protect their privacy than the elderly (Maple, 2017).

The principle of data minimisation is set out in Article 5(1)(c)—per personal data must be adequate, relevant and limited to what is necessary for the purposes for which it is processed (Intersoft Consulting, 2019b). Data minimisation therefore fundamentally collides with the concept of Big Data, which involves collecting as much data as possible. In the context of data minimisation, questions arise whether the data is excessive and relevant. Therefore, it is important for organisations to be able to articulate at the outset the need to collect and process specific datasets.

Furthermore, the GDPR states that personal data shall not be retained for longer than necessary after serving the purpose for which the data had been processed (Intersoft Consulting, 2019b); however, this requirement is likely to face challenges in the context of Big Data. The GDPR does not specify the exact timelines for data retention given that they are context-specific (Intersoft Consulting, 2019b) and difficulties that may arise in relation to the storage limitation principle in Big Data analytics. Most importantly storage limitation principle may undermine the predictability of the future as algorithms can potentially compare current data with stored historical data.

The principle of purpose limitation (Intersoft Consulting, 2019f; Intersoft Consulting, 2019g) is seen as a challenge to Big Data and a barrier to the development of Big Data analytics in the absence of clarity of the purpose for which the data will be used. It was observed that the purpose limitation principle restricts the freedom of the organisation needs to collect data for big data analytics to make discoveries and innovations happen.

A privacy impact assessment (Intersoft Consulting, 2019h) is also an important method that can help identify and mitigate privacy risks prior to the processing of personal data in any Big Data scenario. The unique features of Big Data can make some aspects of a privacy impact assessment additionally difficult, but these challenges can be overcome. The impact assessment of complex data collection and processing systems should be conducted by a third party under the supervision of national data protection authorities that define the professional requirements of these third parties to produce unbiased, high standard outcomes (Mantelero & Vaciago, 2015).

Considering the potential challenges, privacy remains a significant concern in the IoT. Therefore, it is important for the service providers to maintain trustworthiness by honouring privacy of the consumer. That is a consumer-friendly essential to allay public fears when adopting new technology. Research suggests there will be 75 billion Internet-connected de vices, in homes around the world by the end of 2025 (Department for Digital, Culture, Media & Sport, National Cyber Security Centre & Warman, 2020). The individuals are likely to be unaware of the processing of their personal data collected using IoT applications. There are only a few IoT-related policies and regulatory frameworks currently in place; therefore, an effective law implementation mechanism is required to protect millions of users who will otherwise fall victims to cyber-related threats and hacks linked to Internet-connected household items. Table 1 provides a summary of identified challenges and proposed solutions.

Table 1 Identified challenges and proposed solutions

Challenges	Proposed solutions
Collection of data from diverse sources, without careful verification of the relevance or accuracy (Mortier et al., 2021)	Use AI technologies to verify the accuracy of collected data
Big Data techniques allow organisations to combine different datasets, and that increases the likelihood of data being capable of identifying living individuals (Brogan, 2019)	Use of a wide range of technical measures, such as data masking, anonymisation, pseudonymisation, aggregation as well as legal and organisational safeguards (Information Commissioner's Office, n.d.-a)
Limited transparency in how data is processed (Information Commissioner's Office, n.d.-a)	Improve transparency by providing privacy notices (Information Commissioner's Office, 2020) and obtaining consent (Intersoft Consulting, 2019e) before processing any collected data
The complexity of Big Data analytics can lead to mistrust (Richard, n.d.)	Improve transparency by providing privacy notices (Information Commissioner's Office, 2020) and obtaining consent (Intersoft Consulting, 2019e) before processing any collected data
The challenge of determining which purposes are compatible with the purpose for which the data was originally collected	Purpose limitation (Intersoft Consulting, 2019f, g). If an organisation has collected personal data for one purpose and then decided to start analysing it for completely different purposes, then the users need to be made aware of the changes, and, where necessary, further consent needs to be obtained
The custodians of stored data may need to retain them for use over a long period for use in the future	Use of technical measures, such as anonymisation and pseudonymisation (Information Commissioner's Office, n.d.-a)
Any breaches will compromise the privacy of the users and have a negative impact on the image of the custodian, affecting trustworthiness, and the users will lose faith not only in the organisation but also in the system	Use of technical measures, such as anonymisation, pseudonymisation, data masking, encryption keys and blockchain technology. Physical security systems such as access control, use of video surveillance and security logs can also be used
Protection of privacy of individuals	Conducting privacy risk assessment will provide an early warning system to detect privacy problems (Intersoft Consulting, 2019h)
Lack of IoT-related policies and regulatory frameworks at the national, regional and global level	It is important to bring countries, multinational organisations, industrial partners, security and IoT specialists from the industry and academia to build dialogues on how to protect personal information generated through IoT. That will enable us to get a balanced view to move forward in developing policies and regulations associated with Big Data
Principles in national and regional laws contradict with advancement of technologies	It is important to review the policies at least twice a year to make sure there is a balance between upcoming technologies and legal mechanisms to protect the privacy of individuals and national security

2.3 Emerging Laws and Regulations of Data Protection in IoT

Legal regulation is of increasing importance for Big Data, particularly for data protection. In this context, the application of established and developing data protection techniques is rapidly evolving. The managing of compliance with the GDPR will play an essential part in the Big Data handling projects involving data harvested from the expanding range of available digital sources. Many organisations do have established data protection governance structures and policy and compliance frameworks in place, and these act as pathfinders towards Big Data governance.

The GDPR has recognised the rapid technological developments and globalisation with a special reference to Big Data technology (Intersoft Consulting, 2019i); therefore, it has provided further opportunity for regulators and organisations to consider Big Data compliance. In particular, the GDPR has introduced specific tools, like privacy by design (Intersoft Consulting, 2019j) and pseudonymisation (Intersoft Consulting, 2019k), to help deal with Big Data. Consequently, the ICO (Information Commissioner's Office, 2019) and other data protection authorities have been addressing Big Data for some time by further developing existing tools like notice and consent, anonymisation and privacy impact assessments in line with GDPR (Information Commissioner's Office, 2019).

The Government of the UK recently launched a consultation process for regulating consumer Internet of Things (IoT) security, the UK will be one of the first countries to legislate specifically in relation to IoT security and other countries are likely to follow the UK model (Beverley-Smith et al., 2020). The UK government has proposed designating a regulator to monitor industry compliance. The proposals included civil enforcement powers, such as fines potentially up to 4% of annual worldwide turnover and product forfeiture, suspension and recall. However, the omission of Wi-Fi security, as has been reported, would have a significant impact on general IoT security (Beverley-Smith et al., 2020).

The EU Cybersecurity Act 2019 initiated the development of a comprehensive cybersecurity certification schemes across the EU, but the USA has so far failed to pass any federal legislation that will match the UK proposal (Beverley-Smith et al., 2020). The Government of UK is engaged with international partners to ensure that the guidelines drive a consistent, global approach to IoT security. As a step forward, in February 2019, ETSI, a global standards organisation, published the first globally applicable industry standard consumer IoT security, based on the UK Government's Code of Practice (Department for Digital, Culture, Media & Sport, National Cyber Security Centre & Warman, 2020).

The UK government introduced a self-regulatory Code of Practice in October 2018 (CoP) and proposed to widen IoT devices related requirements, which included a ban on universal default passwords in IoT products, implementation of the vulnerability disclosure policy and provision of a defined support period in terms of receiving security updates (Beverley-Smith et al., 2020). The proposals covered both producers and distributors, and the intended purpose was for all IoT devices sold in the UK to be compliant with the security requirements, including goods imported from elsewhere (Beverley-Smith et al., 2020). The included obligations were to ensure that all IoT devices met the security requirements, maintain thorough records of compliance and cooperate fully with the regulator.

In January 2020, the UK government announced it was going to introduce new mandatory requirements for IoT device manufacturers for the purpose of improving consumer data security (Fernandez, 2020). The aim was to ensure these products had strong cybersecurity built-in by design and move responsibility to secure their own devices away from the consumers (Fernandez, 2020). The three main requirements included were unique passwords compulsory for all connected devices, provision of a point of contact for the public to report vulnerabilities and a minimum period of security updates specified when sold (Fernandez, 2020).

In places where devices and services process personal data, the custodian should do so in accordance with applicable data protection law, such as the General Data Protection Regulation (GDPR). The emphasis should be for the individuals to remain in control of their personal data that are collected through IoT. In real circumstances, obtaining consent from the users may not be easy. Therefore, the device manufacturers and IoT service providers should make users aware of the way their data is being used, by whom, for what purposes and clear instructions on how to delete their personal data for each device and service (Intersoft Consulting, 2019b). In cases where the data is being kept for a longer period than needed (Intersoft Consulting, 2019b), all the credentials should be stored securely within services and on devices by using techniques like cryptographic keys, device identifiers and initialisation vectors (Department for Digital, Culture, Media & Sport, 2018). In addition, significant sanctions for violations of data protection obligations should be introduced, and mandatory personal data breach notifications should be extended to all areas of personal data processing (Intersoft Consulting, 2019l).

To ensure the implementation of data protection legislation by professionals, the role of data protection officers should be mandatory (Intersoft Consulting, 2019m). In addition to ensuring a high level of compliance, data protection officers themselves can provide data protection education to staff and management of their respective companies. Therefore, they could play an important role in the design of IoT systems by sharing their expert knowledge on data protection with relevant actors.

The proposals seek to protect the privacy of consumers and online security. The emphasis is also on the urgent need to ensure strong cybersecurity built into smart products by design. According to the director of marketing, the concerns over weak IoT security act as a barrier to the delivery of real benefits to individuals and societies (Department for Digital, Culture, Media & Sport, National Cyber Security Centre & Warman, 2020). Therefore, techUK has been supporting the government's commitment to legislate for integrating cybersecurity into consumer IoT products at the design stage (Muncaster, 2020).

2.4 Policies and Standards Landscape for IoT

The data protection aspects of Big Data have been addressed in a number of reports, guidance and policy documents issued at the national and international level over the past few years (Table 2). The report sign posted Big Data's direction of travel

Table 2 Implemented mechanisms and their purposes

Mechanisms	Purposes
The UK government 2013 strategy paper—Seizing the data opportunity: a strategy for UK data capability (Government of UK, 2013)	It planned to address privacy and data protection issues through a clear and pragmatic policy that ensures public trust in the confidentiality of their data while increasing the availability of data to maximise its economic and social value (Government of UK, 2013)
The Executive Office of the US President's May 2014 report—Big Data: Seizing Opportunities, Preserving Value (Government of US, 2015)	This report considered Big Data and privacy both in the public and the private sectors and concluded that the existing privacy notice and consent approach to data privacy may have to be reviewed in the light of Big Data (Government of US, 2015)
The European Commission's 2014 Communication publication—Towards A Thriving Data-Driven Economy (European Commission, 2014)	The report states that policies on issues relevant to Big Data like data protection and security should lead to more regulatory certainty for businesses and create consumer trust in data technologies (European Commission, 2014)
The European Data Protection Supervisor's 2015 (European Union, 2015b)	The EDPS 2015 emphasised that data protection law must continue to protect the existing rights and values even in the context of Big Data (European Union, 2015b)
In March 2017, the ICO published an updated paper on Big Data, artificial intelligence, machine learning and data protection with GDPR compliance element (Information Commissioner's Office, n.d.-b)	This updated paper presents six recommendations to help organisations achieve compliance which include anonymisation, privacy impact assessments (PIAs), appropriate privacy notices, privacy by design, the development of ethical principles and auditable machine learning algorithms (Information Commissioner's Office, n.d.-b)
Use of encryption keys	The practicality of using public key encryption (PKE) for encryption of data also enables decryption using a private key by the recipient, without undermining privacy and security (Pandey et al., 2018)
Implementation of physical security systems	Physical security systems have the capacity to deny data centre access to strangers or staff members, restricted to their status (Rahfaldt, 2020)

and articulated a focus on data solutions and Big Data as a key IT driver over the next two decades (Richard, n.d.).

The UK government 2013 strategy paper—Seizing the data opportunity: a strategy for UK data capability—presented a positive view of the UK's ability to seize the data opportunity (Government of UK, 2013). It addressed privacy and data protection issues through a clear and pragmatic policy to ensure public trust in the confidentiality of their data while increasing the availability of data to maximise its economic and social value (Government of UK, 2013).

The Executive Office of the US President's May 2014 report—Big Data: Seizing Opportunities, Preserving Value (Government of US, 2015)—focused on the way in which Big Data will transform everyday life, and it considered Big Data and privacy both in the public and the private sectors and concluded that the existing notice and consent approach to data privacy may have to be reviewed in the light of Big Data (Government of US, 2015).

The European Commission's 2014 Communication publication—Towards A Thriving Data-Driven Economy (European Commission, 2014)—sets out a number of activities it considered necessary for the EU to be able to seize Big Data opportunities. This report includes a data-friendly legal framework and policies. The report states that policies on issues relevant to Big Data like data protection and security should lead to more regulatory certainty for businesses and create consumer trust in data technologies (European Commission, 2014).

The European Data Protection Supervisor's 2014 (European Union, 2015a) and European Data Protection Supervisor's 2015 (European Union, 2015b) opinion on the challenges of Big Data. The EDPS 2015 emphasised that data protection law must continue to protect the existing rights and values even in the context of Big Data (European Union, 2015b). In general, the EDPS has called on the EU institutions to use the reform of the EU data protection framework to strengthen the data protection mechanisms to protect personal privacy and secure personal information (Richard, n.d.).

In March 2017, the ICO published an updated paper on Big Data, artificial intelligence, machine learning and data protection with GDPR compliance elements (Information Commissioner's Office, n.d.-b). This updated paper refers to the GDPR where relevant, but it is not intended to be a guide to the GDPR. In particular, the ICO presents six recommendations to help organisations achieve compliance which includes anonymisation, privacy impact assessments (PIAs), appropriate privacy notices, privacy by design, the development of ethical principles and auditable machine learning algorithms (Information Commissioner's Office, n.d.-b).

Big Data cannot be secured by way of policies and legal mechanisms only. The use of encryption keys is one effective way to protect Big Data. The practicality of using public key encryption (PKE) for encryption of data also enables decryption using the private key by the recipient, without undermining privacy and security (Pandey et al., 2018). Physical security systems, on the other hand, have built in the capacity to deny data centre access to strangers or staff members, restricted to their status (Rahfaldt, 2020). Similarly, the use of video surveillance and security logs will serve the same purpose (Rahfaldt, 2020). These methods will contribute to maintaining and preserving confidentiality, integrity and generated data availability.

Companies should continually monitor, identify and rectify security vulnerabilities in their own products and services as a part of the product security lifecycle (Department for Digital, Culture, Media & Sport, 2018). On identifying any disclosed vulnerabilities, prompt action should be taken on the organisations. The sharing of known or identified vulnerabilities with the industrial entities will enable them to be best prepared for potential vulnerabilities in the future Internet.

In the absence of any regulation, it is unlikely that privacy, data protection and information security will be addressed meaningfully and adequately by the market. In developing, accepting and implementing policies associated with IoT, careful consideration should be given to avoiding violation of human identity, human integrity, human rights and the privacy of the individual and the public. The control of personal data should remain in their hands. To ensure harmonisation of privacy to a high standard, data protection and information security, the development of a binding global data protection framework for IoT is appropriate and desirable.

3   Security Challenges and Opportunities for IoT Solutions

The Internet of Everything (IoE) is the next step to IoT as it will connect data, processes, devices and people via the Internet (Kalyani & Sharma, 2015). The frog-leap in these exciting technological advancements comes with risks, challenges and opportunities of their own. Most of these risks are security-relevant issues that will have a significant impact on individuals, organisations and governments in general. This section highlights a multitude of IoT security challenges and the proposed solutions.

3.1   Security Challenges

Due to device differences, protocols and services in IoT, there needs to be a set of standards and well-defined architecture with interfaces, data models and protocols. There is a concern that many researchers are focused mainly on authentication and access control protocols. When IoT devices are connected for the first time and share identifying information, many attacks can happen such as the man-in-the-middle (MITM) attack. To this end, authors in (Mahmoud et al., 2016) stated that cryptography applied by predefined identity management entities that can monitor the connection of devices is needed to prevent identity theft. IoT requires more devices that will switch the use from IPv4 to IPv6 which will require more bandwidth. The implementation of both IPv6 and 5G and the new generation of communication for better speed also open the doors for more threats and challenges that need to be addressed.

Different features of IoT devices can create threats and security challenges (Zhou et al., 2019). A better understanding of these features can help us mitigate some of these issues and rely on the consequences for a better solution. Features such as mobility, interdependency, diversity, intimacy and many more bring different challenges and threats such as firmware vulnerabilities, storage, computing power, network attacks, policies and standards that require more research. It requires thorough investigations to identify the root causes of IoT threats and also to build pragmatic countermeasures (e.g. 'the real risk which may be involved behind these vulnerabilities in the industrial context needs further investigation in the future' (Varga et al., 2017).

There are methods that use blockchain to ensure privacy and security (Dorri et al., 2017). Confidentiality, authorisation, integrity and availability are achieved by using symmetric encryption, shared keys, hashing and limiting acceptable transactions by the device. This method could be manageable for low-resource IoT devices; however, it produces some delay. The delay and the extra overheads are insignificant compared to its security and privacy gains to some applications but critical in others. Also, there is a blockchain IoT system that manages keys using RSA public key (Huh et al., 2017). In this work, private keys are stored in the devices, and public keys are stored in Ethereum. The proposed idea was implemented in a small-scale IoT system, and only a few IoT devices were used. The system showed two weaknesses. The first is the time it requires for data transactions, and the latter is the requirement for larger storage for light IoT devices. In terms of threat and security, prevention from DDoS attacks was the only mentioned security measure that the system could provide. Data encryption is used to limit security risks as they increase for both business and consumers in the IoT environment, and studies show that using AES in the algorithm is faster than both HAN and RSA algorithms (Yousefi & Jameii, 2017).

There are major forensic challenges that face the IoT domain as there is no reliable and documented tool to collect residual evidence (Conti et al., 2018). The autonomous and real-time interactions with different IoT devices and nods make it difficult to collect, identify and preserve evidence data. Identifying activities of different parties that can access IoT nods is a challenge with the lack of a proper authentication system.

As there are some solutions that can be implemented to mitigate the security concerns, 'there is a clear lack of performance evaluation and assessment in real-life scenarios. Furthermore, there is a conflict between protecting user privacy and the granularity of data access needed to provide better services. This raises the challenge of how to support consumer-specific privacy preferences while maintaining the same level of service' (Seliem et al., 2018).

3.2   Proposed Secure IoT Architectures

There is no single architecture or model of IoT. The proposed layer models vary from a three-layer model to a six-layer model. Many technologies are involved to create an IoT system such as RFID, WSN, cloud computing and different network technologies. This may result in different IoT security and privacy challenges such as unauthorised access to RFID, sensor-nodes security breach and cloud computing abuse.

To mitigate the threats that the IoT technology faces, there should be a better understanding of the technology used, architecture, type of attacks and where they all meet.

Different used IoT layering systems are as follows: the three-layer approach is used by (Mahmoud et al., 2016; Seliem et al., 2018; Jia et al., 2012) (application, network and perception layer). The four-layer approach is used by (Varga et al., 2017) (application, data processing, network and sensor and actuators layer). The four-layer approach used by (Farooq et al., 2015a; Leloglu, 2017) (application, middleware, network and perception layer). The six-layer approach is used by (Farooq et al., 2015b) (business, application, middleware, network, perception and coding layer). The three layer approach is used by (Yousefi & Jameii, 2017; Conti et al., 2018) (application, transport and sensing layer).

Many studies present the threats and challenges within IoT based on a layering system faces. There are different layering approaches which make it difficult to allocate the same problem from one layering system to another. This increases the complexity and the time needed to find a proper solution. Here, we used the simplest layering system (Fig. 3) to demonstrate the most essential factors in a simpler way.

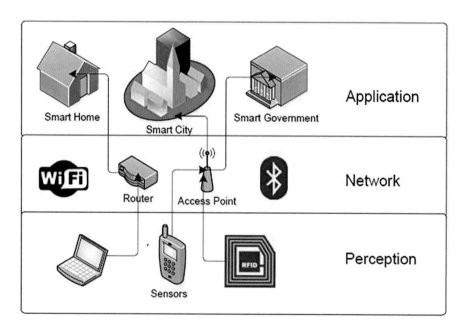

Fig. 3 IoT layers (Mahmoud et al., 2016)

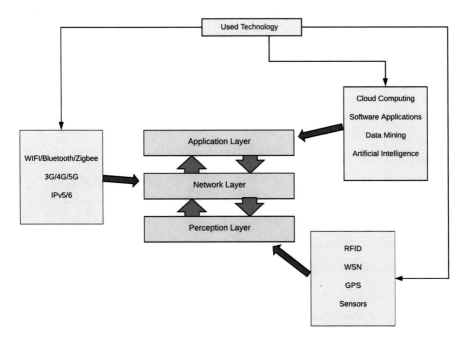

Fig. 4 Used technologies in IoT layers

First, the authors describe the most important technologies used in each layer bearing in mind that technology can be used in more than one layer. Figure 4 provides a simplified example of some technologies used.

In Fig. 4, the authors explain different technologies used in each layer. Threats can then be divided by the technology used rather than the layers they are in. This enabled authors to focus on the main technology used and how to implement the appropriate method to mitigate threats.

Figure 5 demonstrates risks associated with the used technology. This enables threats to be identified with ease. IoT systems do not facilitate all the technologies at once; thus not all protection methods should be implemented. Protection and mitigation methods should be implemented based on the technology used. An example of this would be a system that uses either Bluetooth or Zigbee technology. Security implementation can be specific for the technology used rather than for all the options. This is very important for lightweight IoT devices because protection and security mechanisms tend to need more storage resources and computing power (Zhou et al., 2019; Varga et al., 2017; Huh et al., 2017).

An example of this would be listing the technologies that would be used in IoT rather than the layers in the model, creating a manual or a table (such as Table 3) that lists all the used technologies, their threats and mitigation methods. In a simplified scenario, a company may need to create a new IoT device/application to serve a specific purpose. Users or researchers could first check all the technologies that will be used to create this tool (e.g. not all the devices require cloud computing

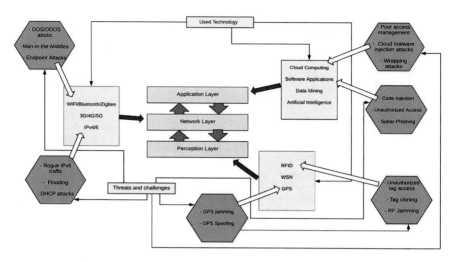

Fig. 5 Threats on the used technologies

Table 3 Example of used technologies and their implementations

IoT technologies	Used/not used	Threats	Implementation
GPS	☐	• GPS jamming • GPS spoofing	• Implement blocking antennas • Obscure antennas
IPv4	☐	• DHCP flooding	• Implement port security
Cloud computing	☐	• Interception of data	• Advanced web application firewalls • Data encryption

technology). Therefore, after an initial evaluation of the used technology, the appropriate control measures can be added to mitigate the threats associated with the technology. For instance, if GPS technology is used in the device, the 'blocking antennas' method can be implemented as a control measure (refer to Table 3).

4 Future Privacy and Security Landscape of IoT (Post-COVID-19)

COVID-19 has made people work from home, shop online and students learn online. It is envisaged that these new normalcies will remain post COVID-19 as well. There are many privacy and security challenges associated with this new normalcy. Such challenges are not a phenomenon unique to the context of COVID-19. Yet both cyber threats and the enforcement gap were running at unacceptably high levels before the pandemic and have continued to do so throughout the crisis (Peters, 2021).

Even though the long-term impact of the COVID-19 crisis on the evolving threat of cybercrime cannot yet be assessed, there are several pressing questions about how the developments seen during the pandemic will affect the future privacy and security of people. Policymakers, practitioners and advocates will have to come up with mandatory risk assessment frameworks to make sure the technology development companies will follow a strict risk assessment before they deploy any innovative technologies. This will prevent any security and privacy complications in the near future.

The response of the government and the technology industries to the coronavirus outbreak became headlines news, but at the same time, concerns were raised about the contact tracing apps, mobile location data tracking and police surveillance drones (Holmes et al., n.d.). Also, new privacy issues have emerged as the organisations started strengthening surveillance using thermal cameras and face-recognition technology in preparation for the resumption of normal working patterns. At one point during the pandemic, the WHO called the situation an infodemic due to the increased collection of information (WHO, UN, UNICEF, UNDP, UNESCO, UNAIDS, ITU, 2020). According to the findings released from a survey conducted in the USA, more than two-thirds of respondents believe that their government should be able to bring the virus under control without them having to sacrifice their privacy (Lovejoy, 2020). In this context, the governments, having to comply with the use of surveillance tools in combating the pandemic, should also need to strike a balance without compromising data privacy laws.

In a post-COVID-19 world, we cannot expect the world to behave comparatively in the same manner as it did in pre-COVID-19. It is extremely necessary to address privacy and security concerns during and in post-COVID-19. In doing so, the private sector can play an effective role in identifying cybercriminals and avoid disruptions to their infrastructure, but only the governments have the legal authority to prosecute and bring them to justice (Daniel et al., n.d.). Therefore, it is crucially important for the public and private sectors to work together on cybercrime issues. That having said, the possibility of some disparities in organisational culture and capacity between the institutions cannot be discounted.

As it stands, there is a clear visible gap in the development of IoT devices, and regulatory laws do exist. Therefore, it is imperative to revisit national and regional data protection mechanisms to address upcoming potential threats, and it would be beneficial to capture data protection principles highlighted in the GDPR. The specific principles such as anonymisation, pseudonymisation, right to be erasure, obtaining consent before collecting and processing of personal information, deletion of collected data within a specified time scale and informing the data subject how the organisations will use their personal information. The adherence to these principles helps build a trustworthy relationship between data controllers and the data subject. However, some have opined that revisiting data protection laws and regulations such as GDPR will jeopardise the success of Big Data (Zarsky, 2017; Bentotahewa & Hewage, n.d.).

5   Conclusion

This chapter discusses the process of Big Data generated through IoT, the challenges and opportunities that have come to light during the COVID-19 pandemic. The authors have reviewed the conceptual meaning of 'BIG Data' and the process of generating a vast volume of data as the definition suggests. The nations have relied on technological solutions to minimise and contain the spread of the pandemic, and the increase in numbers of IoTs connected through the Internet has generated a vast volume of information. As much as the outcomes are tangible and clearly visible, the focus has shifted to concerning security implications on personal privacy and security. In searching for solutions, the authors have identified the importance of accepting and implementing laws, regulations and policies associated with IoT, with a special focus on GDPR. In this article, the authors have explored legal mechanisms already in place and have highlighted the importance of developing and revisiting national and regional data protection mechanisms. A consensus-based set of legislation in line with data protection principles highlighted in the GDPR is needed to confront future threats against personal privacy and security. Implementation of such policies and technical solutions will provide guidance and binding responsibility on the part of the manufacturers and organisations to protect the privacy of the individual while achieving the objectives of IoT deployment.

References

(COVID-19) Coronavirus restrictions: What you can and cannot do—GOV.UK. (2021). Retrieved April 22, 2021, from https://www.gov.uk/guidance/covid-19-coronavirus-restrictions-what-you-can-and-cannot-do#businesses-and-venues

Altuntas, F., & Gok, M. S. (2021). The effect of COVID-19 pandemic on domestic tourism: A DEMATEL method analysis on quarantine decisions. *International Journal of Hospitality Management, 92*, 102719.

Bentotahewa, S., & Hewage, C. (n.d.). *Challenges and obstacles to application of GDPR to Big Data*. Info Security. Retrieved January 23, 2021, from https://www.infosecurity-magazine.com/next-gen-infosec/challenges-gdpr-big-data/

Berkay Celik, Z., Fernandes, E., Pauley, E., Tan, G., & McDaniel, P. (2019). Program analysis of commodity IoT applications for security and privacy: Challenges and opportunities. *ACM Computing Surveys, 52*(4), 74. 30 pages. https://doi.org/10.1145/3333501

Bertino, E. (2016, March). Data security and privacy in the IoT. In *EDBT* (pp. 1–3).

Beverley-Smith, H., Perowne, C. H. N., & Weiss, J. G. (2020). Internet of Things: How the U.K.'s regulatory plans could raise compliance standards. *The National Law Review, XI*(104) Retrieved from https://www.natlawre-view.com/article/internet-things-how-uk-s-regulatory-plans-could-raise-compliance-standards

Brogan, C. (2019). *Anonymising personal data 'not enough to protect privacy', shows new study*. Imperial College. Retrieved November 15, 2019, from https://www.imperial.ac.uk/news/191112/anonymising-personal-data-enough-protect-privacy/

Conti, M., Dehghantanha, A., Franke, K., & Watson, S. (2018). Internet of Things security and forensics: Challenges and opportunities. *Future Generation Computer Systems, 78*.

COVID-19. (2021). Retrieved April 22, 2021, from https://apha.org/Topics-and-Issues/Communicable-Disease/Coronavirus

COVID-19 and the retail sector: Impact and policy responses. (2020, June 16). Retrieved April 23, 2021, from https://www.oecd.org/coronavirus/policy-responses/covid-19-and-the-retail-sector-impact-and-policy-responses-371d7599/

Covid-19 Temperature Screening Service & Test | Metro Security. (2021). Retrieved April 23, 2021, from https://www.metrosecurity.co.uk/services/temperature-screening/

Daniel, M., et al. (n.d.). *How do we beat COVID-19 cybercrime? By working together*. World Economic Forum. Retrieved January 3, 2021, from https://www.weforum.org/agenda/2020/07/alliance-tackling-covidclass="-No-break">-19-cybercrime

Department for Digital, Culture, Media & Sport. (2018). Code of Practice for Consumer IoT Security, United Kingdom. Retrieved from https://assets.publishing.service.gov.uk/government/uploads/system/uploads/attachment_data/file/971440/Code_of_Practice_for_Consumer_IoT_Security_October_2018_V2.pdf

Department for Digital, Culture, Media & Sport, National Cyber Security Centre, & Warman, M. (2020) Government to strengthen security of Internet-connected products. Retrieved May 14, 2020, from https://www.gov.uk/government/news/government-to-strengthen-security-of-internet-connected-products

Dorri, A., Kanhere, S. S., Jurdak, R., & Gauravaram, P. (2017). Blockchain for IoT security and privacy: The case study of a smart home. In *2017 IEEE International Conference on Pervasive Computing and Communications Workshops, PerCom Workshops 2017.*

Dutta, P. K., & Mitra, S. (2021). Application of agricultural drones and IoT to understand food supply chain during post COVID-19. *Agricultural Informatics, 2021,* 67–87.

Embree, R. (2021, March 19). Four IoT trends for hospitality | Hospitality Technology. Retrieved April 22, 2021, from https://hospitalitytech.com/four-iot-trends-hospitality

Erevelles, S., Fukawa, N., & Swayne, L. (2016). Big Data consumer analytics and the transformation of marketing. *Journal of Business Research, 69*(2), 897–904. https://doi.org/10.1016/j.jbusres.2015.07.001

European Commission. (2014). In European Commission (Ed.), *Towards a thriving data-driven economy*. Retrieved from https://ec.europa.eu/transparency/regdoc/rep/1/2014/EN/1-2014-442-EN-F1-1.Pdf

European Union. (2015a). *European Data Protection Supervisor—Resolutions, recommendations and opinions*. European Union. Retrieved from https://ec.europa.eu/dorie/fileDownload.do;jsessionid=UdwG4bm1A8b_m1-1-UyfY02xUZ1JtAlxTYCJelGukIsnFGJyQCuC!-898031139?docId=2199637&cardId=2199636

European Union. (2015b). *European Data Protection Supervisor; Annual Report 2015*. European Union. Retrieved from https://ec.europa.eu/dorie/fileDownload.do;jsessionid=UdwG4bm1A8b_m1-1-UyfY02xUZ1JtAlxTYCJelGukIsnFGJyQCuC!-898031139?docId=2199637&cardId=2199636

Farooq, M. U., Waseem, M., Khairi, A., & Mazhar, S. (2015a). A critical analysis on the security concerns of Internet of Things (IoT). *International Journal of Computer Applications, 111*(7).

Farooq, M. U., Waseem, M., Mazhar, S., Khairi, A., & Kamal, T. (2015b). A review on Internet of Things (IoT). *International Journal of Computer Applications, 113*(1).

Fernandez, A. (2020). New IoT security regulations: What you need to know. Retrieved April 12, 2020, from https://www.allot.com/blog/new-iot-security-regulations-what-you-need-to-know/

Government of UK. (2013). *Seizing the data Opportunity; A strategy for UK data capability*. Government Publication. Retrieved from https://assets.publishing.service.gov.uk/government/uploads/system/uploads/attachment_data/file/254136/bis-13-1250-strategy-for-uk-data-capability-v4.pdf

Government of US. (2015). *Big Data: Seizing opportunities, preserving values*. Government Publication. Retrieved from https://obamawhitehouse.archives.gov/sites/default/files/docs/20150204_Big_Data_Seizing_Opportunities_Preserving_Values_Memo.pdf

Holmes, O., McCurry, J., & Safi, M. (n.d.). Coronavirus mass surveillance could be here to stay, experts say. *The Guardian*. Retrieved February 4, 2021, from https://www.theguard-ian.com/world/2020/jun/18/coronavirus-mass-surveillance-could-be-here-to-stay-tracking

Huh, S., Cho, S., & Kim, S. (2017). Managing IoT devices using blockchain platform. In *International Conference on Advanced Communication Technology, ICACT*.

Ilieva, G., & Yankova, T. (2020). IoT in distance learning during the COVID-19 pandemic. *TEM Journal, 9*(4).

Information Commissioner's Office. (2019). The UK GDPR. Retrieved November 24, 2019, from https://ico.org.uk/for-organisations/dp-at-the-end-of-the-transition-period/data-protection-now-the-transition-period-has-ended/the-gdpr

Information Commissioner's Office. (2020). What privacy information should we provide?. Retrieved March 17, 2020, from https://ico.org.uk/for-organisations/guide-to-data-protection/guide-to-the-general-data-protection-regulation-gdpr/the-right-to-be-informed/what-privacy-information-should-we-provide/

Information Commissioner's Office. (n.d.-a). Big data and data protection. Version 1.0 (pp. 12–13). Retrieved from https://rm.coe.int/big-data-and-data-protection-ico-information-commissioner-s-office/1680591220

Information Commissioner's Office. (n.d.-b). Big data, artificial intelligence, machine learning and data protection. Version 2.2 (p. 59). Retrieved from https://ico.org.uk/media/for-organisations/documents/2013559/big-data-ai-ml-and-data-protection.pdf

Intersoft Consulting. (2019a). Art. 4 GDPR Definitions. General Data Protection Regulation (GDPR). Retrieved June 4, 2019, from https://gdpr-info.eu/art-4-gdpr/

Intersoft Consulting. (2019b). Art. 5 GDPR Principles relating to processing of personal data. General Data Protection Regulation (GDPR). Retrieved June 12, 2019, from https://gdpr-info.eu/art-5-gdpr/

Intersoft Consulting. (2019c). Art. 3 GDPR Territorial scope. General Data Protection Regulation (GDPR). Retrieved June 16, 2019, from https://gdpr-info.eu/art-3-gdpr/

Intersoft Consulting. (2019d). Recital 26—Not applicable to anonymous data*. Recitals. Retrieved May 26, 2019, from https://gdpr-info.eu/recitals/no-26/

Intersoft Consulting. (2019e). GDPR consent. General Data Protection Regulation (GDPR). Retrieved May 20, 2019, from https://gdpr-info.eu/issues/consent/

Intersoft Consulting. (2019f). Art. 9 GDPR processing of special categories of personal data. General Data Protection Regulation (GDPR). Retrieved September 10, 2019, from https://gdpr-info.eu/art-9-gdpr/

Intersoft Consulting. (2019g). Chapter 9: Provisions relating to specific processing situations. General Data Protection Regulation (GDPR). Retrieved September 15, 2019, from https://gdpr-info.eu/chapter-9/

Intersoft Consulting. (2019h). GDPR Privacy Impact Assessment. General Data Protection Regulation (GDPR). Retrieved June 2, 2019, from https://gdpr-info.eu/issues/privacy-impact-assessment/

Intersoft Consulting. (2019i). Recital 6—Ensuring a high level of data protection despite the increased exchange of data*. Recital. Retrieved April 2, 2019, from https://gdpr-info.eu/recitals/no-6/

Intersoft Consulting. (2019j). GDPR Privacy by Design. General Data Protection Regulation (GDPR). Retrieved April 2, 2019, from https://gdpr-info.eu/issues/privacy-by-design/

Intersoft Consulting (2019k). Recital 28—Introduction of pseudonymisation. Recital. Retrieved April 2, 2019, from https://gdpr-info.eu/recitals/no-28/

Intersoft Consulting. (2019l). GDPR fines/penalties. Key issues. Retrieved September 15, 2019, from https://gdpr-info.eu/issues/fines-penalties/

Intersoft Consulting. (2019m). Art. 37 GDPR Designation of the data protection officer. General Data Protection Regulation (GDPR). Retrieved September 15, 2019, from https://gdpr-info.eu/art-37-gdpr/

Jia, X., Feng, Q., Fan, T., & Lei, Q. (2012). RFID technology and its applications in Internet of Things (IoT). In *2012 2nd International Conference on Consumer Electronics, Communications and Networks, CECNet 2012—Proceedings.*

Jurcut, A., Niculcea, T., Ranaweera, P., et al. (2020). Security considerations for Internet of Things: A survey. *SN Computer Science, 1*, 193. https://doi.org/10.1007/s42979-020-00201-3

Kalyani, V. L., & Sharma, D. (2015). IoT: Machine to Machine (M2M), Device to Device (D2D) Internet of Everything (IoE) and Human to Human (H2H): Future of communication. *Journal of Management Engineering and Information Technology, 26.*

Leloglu, E. (2017). A review of security concerns in Internet of Things. *Journal of Computer and Communications, 05*(01).

Lovejoy, K. (2020). *COVID-19: How future investment in cybersecurity will be impacted.* EY Global Consulting. Retrieved October 25, 2020, from https://www.ey.com/en_gl/consulting/how-the-covid-19-pandemic-is-impacting-future-investment-in-security-and-privacy

Mahmoud, R., Yousuf, T., Aloul, F., & Zualkernan, I. (2016). Internet of Things (IoT) security: Current status, challenges and prospective measures. In *2015 10th International Conference for Internet Technology and Secured Transactions, ICITST 2015.*

Mantelero, A., & Vaciago, G. (2015). Data protection in a big data society. Ideas for a future regulation. *Digital Investigation, 15*, 104–109. https://doi.org/10.1016/j.diin.2015.09.006

Maple, C. (2017). Security and privacy in the Internet of Things. *Journal of Cyber Policy, 2*(2), 155–184. https://doi.org/10.1080/23738871.2017.1366536

McDermott, Y. (2017). Conceptualising the right to data protection in an era of Big Data. *Big Data & Society, 4*(1). https://doi.org/10.1177/2053951716686994

Mortier, S., Debussche, J., & César, J. (2021). Big Data & issues & opportunities: Privacy and data protection. *Bird and Bird.* Retrieved January 12, 2021, from https://www.twobirds.com/en/news/articles/2019/global/big-data-and-issues-and-opportunities-privacy-and-data-protection

Muncaster, P. (2020). UK's IoT Law hopes to drive security-by-design. Infosecurity. Retrieved July 23, 2020, from https://www.infosecurity-magazine.com/news/uks-iot-law-hopes-to-drive/

Obaidat, M. A., Obeidat, S., Holst, J., Al Hayajneh, A., & Brown, J. (2020). A comprehensive and systematic survey on the Internet of Things: Security and privacy challenges, security frameworks, enabling technologies, threats, vulnerabilities and countermeasures. *Computers, 9*, 44. https://doi.org/10.3390/computers9020044

Oussous, A., Benjelloun, F., Ait Lahcen, A., & Belfkih, S. (2018). Big Data technologies: A survey. *Journal of King Saud University Computer and Information Sciences, 30*(4), 431–448. https://doi.org/10.1016/j.jksuci.2017.06.001

Panchal, J. (2019, January 7). How IoT-enhanced warehouses are changing the supply chain management—Part 1—IoT Now News— How to run an IoT enabled business. Retrieved March 3, 2021, from https://www.iot-now.com/2019/01/07/91762-iot-enhanced-warehouses-changing-supply-chain-management/

Pandey, K. K., Rammilan, & Shukla, D. (2018). *Security and privacy challenges in Big Data* (pp. 74–77). Researchgate. Retrieved from https://www.research-gate.net/publication/324482789_Security_and_Privacy_Challenges_in_Big_Data

Pesheva, E. (2020). Coronavirus and the heart. Retrieved July 21, 2020, from https://hms.harvard.edu/news/coronavirus-heart

Peters, A. (2021). Is COVID-19 changing the cybercrime landscape? In *The COVID-19 pandemic and trends in technology.* Chatham House. ISBN: 978 1 78413 436 5.

Pillai, S. G., Haldorai, K., Seo, W. S., & Kim, W. G. (2021). COVID-19 and hospitality 5.0: Redefining hospitality operations. *International Journal of Hospitality Management, 94*, 102869.

Rahfaldt, K. (2020). How leveraging big data changes the perception of security. Retrieved February 22, 2020, from https://www.securitymagazine.com/articles/90766-how-leveraging-big-data-changes-the-perception-of-security

Richard, K. (n.d.). Big data and data protection (UK). *Practical Law.* Retrieved March 14, 2021, from https://uk.practicallaw.thomsonreuters.com/w-017-1623?transitionType=Default&context-Data=(sc.Default)&firstPage=true

Rosenthal, J. N., & Oberly, D. J. (2020). The rise of Internet of Things security laws: Part I. *Pratt's Privacy & Cybersecurity Law Report, 6*(5), 155–158. Retrieved from https://www.jdsupra.com/legalnews/the-rise-of-internet-of-things-security-50035/

Rowan, N. J., & Galanakis, C. M. (2020). Unlocking challenges and opportunities presented by COVID-19 pandemic for cross-cutting disruption in agri-food and green deal innovations: Quo Vadis? *Science of the Total Environment, 748*, 141362.

Self Checkout Systems in 2021: Comprehensive Guide. (2021, January 6). Retrieved April 23, 2021, from https://research.aimultiple.com/self-checkout/

Seliem, M., Elgazzar, K., & Khalil, K. (2018). Towards privacy preserving IoT environments: A survey. *Wireless Communications and Mobile Computing, 2018*, 1032761.

Song, H. J., Yeon, J., & Lee, S. (2021). Impact of the COVID-19 pandemic: Evidence from the U.S. restaurant industry. *International Journal of Hospitality Management, 92*, 102702.

Statista. (2021). *Forecast end-user spending on IoT solutions worldwide from 2017 to 2025 (in billion U.S. dollars)*. Statista. Retrieved March 14, 2021, from https://www.statista.com/statistics/976313/global-iot-market-size/

Suleman, H. (2021, March 12). How to use the IoT to keep your restaurant clean and safe | Food- SafetyTech. Retrieved April 22, 2021, from https://foodsafetytech.com/column/how-to-use-the-iot-to-keep-your-restaurant-clean-and-safe/

Tawalbeh, L., Muheidat, F., Tawalbeh, M., & Quwaider, M. (2020). IoT privacy and security: Challenges and solutions. *Applied Sciences, 10*(12), 4102. https://doi.org/10.3390/app10124102

Triax Technologies. (2021). *Proximity trace*. Triax Technologies. Retrieved April 23, 2021, from https://www.triaxtec.com/resource/fact-sheet/proximity-trace/

UK Anonymization Network. (2021). Retrieved March 23, 2021, https://ukanon.net/

Varga, P., Plosz, S., Soos, G., & Hegedus, C. (2017). Security threats and issues in automation IoT. In *IEEE International Workshop on Factory Communication Systems—Proceedings, WFCS*.

W. H. Organization. (2021). Coronavirus. Retrieved April 22, 2021, from https://www.who.int/health-topics/coronavirus#tab=tab_1

Waheed, N., He, X., Ikram, M., Usman, M., Hashmi, S. S., & Usman, M. (2020). Security and privacy in IoT using machine learning and blockchain: Threats and countermeasures. *ACM Computing Surveys, 53*(6), 122. 37 pages. https://doi.org/10.1145/3417987

WHO, UN, UNICEF, UNDP, UNESCO, UNAIDS, ITU. (2020). UN Global Pulse, and IFRC Managing the COVID-19 infodemic: Promoting healthy behaviours and mitigating the harm from misinformation and disinformation. WHO. Retrieved December 18, 2020, from https://www.who.int/news/item/23-09-2020-managing-the-covid-19-infodemic-promoting-healthy-behaviours-and-mitigating-the-harm-from-misinformation-and-disinformation

Wu, X., Zhu, X., Wu, G., & Ding, W. (2014). Data mining with big data. *IEEE Transactions on Knowledge and Data Engineering, 26*(1), 97–107. https://doi.org/10.1109/TKDE.2013.109

Yousefi, A., & Jameii, S. M. (2017). Improving the security of internet of things using encryption algorithms. In *IEEE International Conference on IoT and its Applications, ICIOT 2017*.

Yousif, M., Hewage, C., & Nawaf, L. (2021). IoT Technologies during and beyond COVID-19: A comprehensive review. *Future Internet, 13*, 105. https://doi.org/10.3390/fi13050105

Zarsky, T. Z. (2017). Incompatible: The GDPR in the age of Big Data. Retrieved from https://scholarship.shu.edu/cgi/viewcontent.cgi?article=1606&context=shlr

Zhou, W., Jia, Y., Peng, A., Zhang, Y., & Liu, P. (2019). The effect of IoT new features on security and privacy: New threats, existing solutions, and challenges yet to be solved. *IEEE Internet Things Journal, 6*(2).

The Challenges of the Internet of Things Considering Industrial Control Systems

Kim Smith (iD) **and Ian Wilson** (iD)

1 Introduction

1.1 Internet of Things

There are many authors who have described what the Internet of Things (IoT) is. Author Greengard (2015) introduces the subject of IoT along with multiple articles (Madakam et al., 2015; Khan & Salah, 2018). They present an introduction to the concept of IoT. Authors Madakam, Ramaswamy, and Tripathi (2015) reviewed literature on the IoT concept with the conclusion that there is no common definition of the term. Authors have tried to identify the origins of the terminology. The suggestion by sources (Greengard, 2015; Postscapes, 2020) is that Kevin Ashton, the Executive Director of Auto-ID Labs in MIT in 1999, was the first person to make use of the term IoT. He was at the time working on a presentation for Procter & Gamble in the context of RFID supply chains.

The definition adopted throughout this article will be that provided by the Centre for the Protection of National Infrastructure (CPNI) (Centre for the Protection of National Infrastructure, 2021) which offers a definition that presents a network of devices with autonomous functions which are part of everyday life.

The IoT as described is something that exists everywhere that a connection to the Internet is possible. The connection mechanism does not concern the IoT. As in Miller (2015) any device that can be uniquely identifiable (normally by an IP address) can be considered as a part of the IoT. This is not just devices we consider as digital such as laptops or smart phones but also includes those domestic devices such as washing machines, lights, and heating that can be controlled remotely.

K. Smith (✉) · I. Wilson
University of South Wales, Treforest, UK
e-mail: kim.smith@southwales.ac.uk; Ian.Wilson@southwales.ac.uk

© The Author(s), under exclusive license to Springer Nature Switzerland AG 2022   77
R. Montasari et al. (eds.), *Privacy, Security And Forensics in The Internet of Things (IoT)*, https://doi.org/10.1007/978-3-030-91218-5_4

1.2 Industrial Control System

Multiple authors have described Industrial Control Systems (ICS) in peer-reviewed articles as well as in academic materials. Authors (National Institute of Standards and Technology, 2011; Simon, 2017; Assenza & Setola, 2019; National Institute of Standards and Technology, 2008; Hayden et al., 2014; Bodungen et al., 2017) introduce the modern concept of ICS; however, ICS was first identified in Greek and Arabian societies. The literature sources surrounding ICS use a different terminology that leads to confusion. One form of terminology used to describe an ICS is a Process Control System (PCS). Another terminology used is Supervisory Control and Data Acquisition (SCADA). This describes one of the topologies of ICS. The different topologies of ICS are PCS, SCADA, Distributed Control System (DCS), SMART, or Industrial Automation and Control Systems (IACS).

The definition to be adopted throughout this article will be that provided by the National Institute of Standards and Technology in their Glossary of Terms in NIST SP-800 (National Institute of Standards and Technology, 2011) that describes information systems that control remote assets and local assets utilized in industrial processes including manufacturing, distribution, and other production processes.

2 Industrial Control Systems

ICS are different from IoT, but they are also similar. This section is aimed at providing a more in-depth introduction to ICS and how they are similar to IoT. An ICS is different because it is based on industry and will have a combination of operational and information technology. An IoT will tend to be more based on a residential setting and be based on information technology only. However, current development is presenting the Industrial Internet of Things (IIoT). In his report (Simon, 2017) the author describes the IIoT in terms of the communication that occurs between machines and the immense volumes of data that are generated that can support the development of efficient industry processes.

2.1 Operational Technology

Operational technology (OT) is only relevant in an industry setting. In their article (Assenza & Setola, 2019) the authors define OT as a system with assets that are linked together to monitor and control automated processes through information communication technology.

2.2 Information Technology

Information technology (IT) is a supporting structure for both industry and the citizens of the world. It consists of a diverse range of digital devices from computers to IoT devices such as smart washing machines and heating controls. The other element of IT is the communication media that is used. There are also many forms of media, but they all provide a connection to the Internet whether through Bluetooth, wireless, or Ethernet technology.

2.3 Functions of ICS

A typical ICS operation is described by NIST (National Institute of Standards and Technology, 2008), and the fundamental structure is a closed-loop control system also known as a feedback loop. A closed-loop control system has the primary aim of processing information in the following manner:

- Accept an item of data usually from a sensor.
- Feed the data to a process.
- Perform a process using the data and the feedback data.
- Output an item of data.
- Feed the output data (feedback data) into the process.

This is performed in a cyclic manner as shown in Fig. 1.

This basic principle is embedded into all ICS and is further defined by authors discussing the main functions of ICS. In their SANS whitepaper, Hayden et al. (2014) offer four main functions of an ICS as measure, compare, compute, and correct. NIST supports this in their description of the ICS components and operations (National Institute of Standards and Technology, 2008) in which they define four elements as measure, compare, compute, and correct. In their book Bodungen et al. (2017) consider only three functions of ICS as view, monitor, and control.

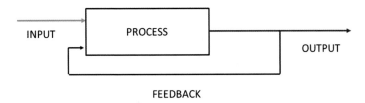

FEEDBACK

Fig. 1 A closed-loop control system

2.4 Physical Components of Industrial Control Systems

The ICS systems are found in all environments in support of everyday life. The functions as described above are performed by the components of the ICS system. The components are varied and depend partially on the topologies of ICS and the industry sector that they are applied to. The topologies are:

- DCS is used in process-based industry such as agriculture, chemical plants, and automobile manufacturing.
- SCADA is used to monitor and control industries such as oil and gas pipelines and electric power grids.
- PLC is a part of a larger configuration within a SCADA or DCS system.
- SMART is used in residential and industry environment.
- Industrial Automation and Control Systems (IACS) is in a small geographic location such as a manufacturing plant.

Authors Knapp and Langill (2015) describe the components of an ICS in a system-wide context. Others take a physical approach such as the one described by Hayden et al. (2014) in their SANS whitepaper. In their paper they identified the following components of ICS:

- Sensors that perform a measurement task
- Transducers that convert a measurement into an electrical signal
- Transmitters that convert and then send the signal
- Controllers that perform processes on input and provides an output
- Final control elements that make a change based on the signal sent to them

2.5 Commonalities Between ICS and IOT

This mixture of definitions of IoT means that it can be interpreted in many ways. In defining how an ICS is a form of IoT, it is necessary to analyze the definitions to identify the common elements. The result of comparing the definitions is the identification of the following commonality:

- Multiple intelligent devices
- Interconnectivity of devices through the Internet
- Enabling the sharing of big data
- Contained within a closed-loop control system
- Autonomous
- Self-monitoring capability

2.6   Other Components of Industrial Control Systems

As a part of an ICS system, the term socio-technical system (STS) is used to describe components including the physical. An STS consists of complex interactions between humans and technical systems. This term was derived from studies undertaken by Trist (1981) on the effects of technology on workers. The results did not always indicate an improvement in efficiency or productivity, linking these to other factors in the working environment not the technology. The original model of STS consisting of the social and technical systems was presented by Bostrom and Heinen (1977). This model develops the concept around four elements, technology, structure, people, and tasks, and is used to indicate the complexity of the interactions between humans and technology. They describe the system as:

- The technical system is concerned with the processes, tasks, and technology needed to transform inputs to outputs.
- The social system is concerned with the attributes of people (e.g., attitudes, skills, values), the relationships among people, reward systems, and authority structures.

This original model was further developed in 2016 by Oosthuizen and Pretorius in their article (Oosthuizen & Pretorius, 2016) where they add an additional environmental dimension. The environment dimension encircles the STS which contains the elements described by Bostrom and Heinen (1977). This additional element was included to represent the concept that the STS was an open system. Open systems are susceptible to external inputs from the environment, thus increasing the complexity. Other authors offer alternative views of STS. Wu et al. (2015) offer a hierarchy to represent the elements of the STS system. The hierarchy is subdivided into three parts:

- Social
- Technical
- Environment

Each of the subdivisions of the hierarchy is scoped individually, and it is not possible to combine them to attain a holistic view. Authors such as Malatji, Von Solms, and Marnewick (2019) in their paper continue to work within the STS model presented by Oosthuizen and Pretorius (2016)) and in their research identified the people element as the weakest link. They identify that there are many reasons why this is the situation. Their emphasis is to try to uncover gaps and to focus on the effectiveness of current security controls to optimize them.

3   Challenges in Industrial Control Systems

There are many challenges relating to ICS, and to explain these, they have been categorized into the following:

- People
- Physical
- Security
- Organization structure

Challenges in ICS are based on the concept of risk. Managing risk is a very important task within any organization. There are many types of business risk; however, this report is concerned with the risk surrounding the use of ICS and concentrates on the element of cyber risk. Cyber risk is a major concern of the board of an organization, and such things as awareness, budget, culture, and priorities may affect the level at which an organization deals with risk. Supporting the board, employees should have an awareness of cybersecurity, but this will be at different knowledge and skills levels. With a lack of knowledge come mistakes and errors which can increase risk. The statistics from Ernst and Young survey (Fig. 2) show that employees are accepted as the most likely cause of risk in a business (Ernst and Young Global Limited, 2020).

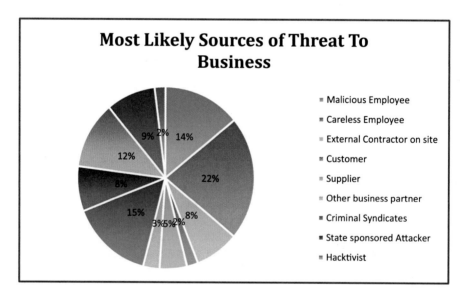

Fig. 2 Most likely sources of threat

3.1 People

People are a key element of an ICS, but they are often ignored in the recognition of the risk level that they give. When considering the human as a part of risk, an important subject is psychology. In terms of ICS and IoT, exploring the area of psychology can help identify certain characteristics and traits that make a person more vulnerable to an attack from social engineering. Three general concepts are:

- Susceptibility
- Awareness
- Motivation

The concept of risk associated with humans relates to different aspects, and authors such as Mouton, Leenen, and Venter (2016) have developed an extensive ontology of attacks, techniques, and other key areas around social engineering. In his book Hadnagy (2011) introduces the concept of social engineering and references definitions from multiple sources. He offers a simple definition in an individual performing an action through the maneuvering by another.

There are many sources of definitions of social engineering. The Oxford University Press states that this is deception by an individual to gather confidential information from another through manipulation.

Developing this along with information from Babu et al. (2017), National Institute of Standards and Technology (2021), and Doan (2006)), the diagrams identifying an ontology of social engineering in Fig. 3 and Fig. 4 demonstrate the complexity of the subject.

Susceptibility

This is concerned with the characteristics and traits of an individual. Individuals develop these traits over time, and a person involved in social engineering is observing in the hope of identifying these traits in support of the development

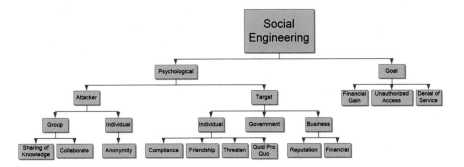

Fig. 3 Social engineering ontology part 1

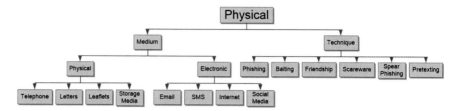

Fig. 4 Social engineering ontology part 2

of an attack. They will watch and observe individuals looking for their habits, routines, and personal behaviors. The Collins dictionary (2016) presents the word susceptibility as the link to the degree to which an individual can be affected by or influenced by another. The habits, routines, and behaviors make individuals a target.

The behaviors that are a clue to a person's susceptibility would be a demonstration of their trust in people. Another could be their integrity; this can be tested by those who are involved in social engineering. Other clues would be a person identifying with their social worth; much of this information can be gathered from social media such as Facebook and Twitter. Other signs will relate to their working environment. The primary needs of an individual as identified by Maslow (2013) should be met particularly the basic needs such as shelter, food, water, and security for them to be less vulnerable.

Awareness

When considering awareness as a contributing factor to social engineering, there are different and conflicting opinions. Awareness can be separated into two key elements. The first is the employee awareness of cybersecurity and the risks and consequences to the organization. This awareness would be a part of a training package for all employees. The second is the awareness of employees of the standard working practices and policies in place to protect the organization from cyber-attacks. These should reduce the risk to the organization. This question of awareness was addressed by (Aldawood et al., 2020) in their article. The article links the security state of a system and the vulnerability of employees. They link people using the most secure systems as often being the most vulnerable to social engineering attacks. This is borne from the false idea that security procedures exist, and employees are aware that they will use them. The reality is that employees will try to find the quickest way to perform a task which could entail the bypassing of the security measures. For example, an employee may receive a USB storage device from a supplier. Procedure should say load into a clean (standalone) pc first; however, the employee trusts the supplier and loads straight onto the network causing malware to be loaded onto the network.

Motivation

Already mentioned is the challenge that employees in a business can be susceptible to a social engineering attack. One consideration is the motivation of the individual in terms of two things, their home life and work life. Employees who are dissatisfied at work have an increased susceptibility to attack. This can be from multiple sources, if an employee has been passed over for promotion or they feel that they are being blamed for things going wrong or even that they did not get a pay rise. These all affect a human's psychological state, and this can be manipulated. The use of social media to vent an individual's frustration is an open door for a social engineer. With motivational factors it is important to remember that this is a person's perspective and may not be true. To enable better security from cyber-attacks, managers must be aware of the human emotional factors of their work force. An article that undertakes a comparison of factors (Alblabi & Weir, 2018) for social engineering provides an analysis of the personal email and social environment which can be crossed into the work environment.

3.2   Physical

The physical challenges of ICS are concerned with the physical components of the ICS. This can be a primary element as described or the communication media of the interconnection between the elements. This chapter will not be used to consider the challenges that relate to the security of such elements as sites as these would be covered under a site management policy. The challenges of the physical components of ICS can be categorized as:

- Legacy
- Maintenance
- Cost
- Commercial off the shelf
- Mitigation of risk

Legacy

ICS are referred to as legacy systems by some authors (Ernst and Young Global Limited, 2020; Kriaa et al., 2019; Ginter, 2016); this happens for several reasons: the age of the system, the lack of vendor support, the older hardware, and an increased cost of maintenance. A simple explanation is provided in Techopedia (2021) defining the system as consisting of outdated components that could be the software, device, or programming language. An important point is that these types of systems were originally in place with the priority to ensure the safety of the system and protection of people and business not the security from attack.

Maintenance

One of the key issues for legacy systems is the subject of maintenance which includes upgrades and patches to software. Kilman and Stamp (2005) identify that many devices in ICS have never been updated with anti-virus or firmware since their installation. There are many reasons why organizations feel that they are unable to perform much-needed maintenance:

- Availability issue and disruption
- Lack of vendor support
- If it is not broken do not fix it attitude
- Too costly
- Not enough skills
- Concerned about the impact to other elements of the system

Babu et al. (2017) support (Kilman & Stamp, 2005) in that a lot of ICS systems have been operational for a long time and therefore are legacy systems. These systems have not been maintained, and the age of the technology implemented means new security options cannot be implemented.

Cost

Cost to a business must be considered for both long term and short term. There could be short-term costs that may give quick results in terms of risk mitigation. However, it is generally considered that the long-term cost is substantial given that devices may have to be upgraded in some way. Cost can also include the mitigation actions. A company may just decide to have devices on standby in case they are attacked. This can be a very expensive option but may be the only possible solution. Having redundant equipment around needs storage and needs to be maintained. The life of the device is a big factor in deciding the cost and the replacement plan.

Commercial Off the Shelf (COTS)

In the NIST glossary (National Institute of Standards and Technology, 2021) describing abbreviations throughout the vast library of NIST documentation, they define COTS as an abbreviation for commercial off the shelf. This means the range of existing software and hardware that is available from commercial outlets. For ICS systems this increases cyber risk which must be mitigated against to ensure minimum risk to life.

In his thesis Dung Doan (2006) introduces the advantages of using COTS items. The advantages are that it incorporates newer technology and newer standards. It can be updated faster than custom-built software. Maintenance cost is substantially reduced since COTS software is widely used by a large population. COTS items although they have advantages also have disadvantages, and the main concern relates to the security of using them. Some typical issues are that COTS software is not amendable, defaults will be in place, easy availability for an attacker, and configuration weaknesses.

COTS software is designed to not be changed and therefore cannot be customized to meet the needs of specific ICS. COTS vendors do not provide any guarantee that the items are secure. Lastly COTS items are designed with functionality as the highest priority; therefore less attention is spent on the security of the software. COTS items have security defaults in place such as administrator overrides. This immediately is a high risk to ICS systems, and all default passwords and user identifiers should be changed as soon as an installation is made. However, many vendors do not provide installers with the information, and so they are not aware of the risk. These provide excellent backdoors for hackers to attack an ICS system.

COTS items are widely available which increases the risk that users with malicious intent can attain them. These users therefore have the potential to uncover security flaws in the items as they take time to analyze how they work. If flaws are identified, there is an increase in risk to the item and the systems that they are embedded in.

The variety of potential risks is wide and putting this into a business context. A study was undertaken by Project SHodan INtelligence Extraction (SHINE) (2014) in 2014. This was a collaboration of organizations and individuals to demonstrate the vulnerability of SCADA systems. Their research demonstrated that there were over one million ICS/SCADA systems connected to the Internet with unique IP addresses. Having identified so many devices, it is easy to select those that are vulnerable and make an attack.

Mitigation of Risk

One type of mitigation of physical risk that is used is defense-in-depth as described by Melissa Tucker (2015) as a multi-layered defense approach. This approach makes use of different cyber-defense mechanism, and this should prevent a single point of failure in the system. This type of strategy is most often used by the military as a complex defense is more difficult and time-consuming to penetrate. This strategy is supported by NASA and other bodies such as the United States Nuclear Regulatory Commission (2016).

In the article written by Kupiers and Fabro (2006), they identify several key differences between traditional IT environments and control system environments and how they affect securing ICS systems. In the article they compare security elements and how they are different between IT and ICS. The comparison identifies the differences in applying patches and anti-virus, the requirements for availability and time criticality of the systems, as well as the lifetime of the components. They also included a comparison of the environment such as outsourcing and the physical situation in remoteness of systems.

The authors after the comparison discuss and identify what they consider as the five key security countermeasures for control systems:

1. Security policies
2. Blocking access to resources and services
3. Detecting malicious activity
4. Mitigating possible attacks
5. Fixing core problems

3.3 Security

ICS systems as well as IoT systems must be secure and safe. In terms of how important they are is based on the evolutional aspect of IT and ICS. Originally safety was the concern of developers of ICS systems minimizing any impact on the environment or ensuring no loss of life or injury. On the other hand, the developers and maintainers of IT systems were originally only concerned with the security. An issue for both areas is that there are different definitions for industry sectors. In their article the authors (Kriaa et al., 2019) define the difference between security and safety:

- Safety—the risk that is accidental but has unacceptable results
- Security—risk that is malicious

Another perspective was given by Andrew Ginter in his book (Ginter, 2016) where he defines cybersecurity as the prevention of attacks and that ICS security is the prevention of unauthorized operation of the system. Author Stig Johnson (Johnson, 2013) discusses resilience-based risk management and offers an alternative description of safety and security. He stated that safety was concerned with the accidental harm prevention, reduction, and reaction to systems. In comparison he stated that security was concerned with malicious harm prevention, reduction, and reaction to systems.

CIA/AIC Triad Model

The CIA triad model is a building block for security policies utilized by organizations. The model is a start point in the understanding of the security of ICS and is utilized by many different industries. There are three factors of the model: integrity, availability, and confidentiality.

- Confidentiality—is concerned with the protection of personal data, and its loss can have a huge impact on an organization both financially and reputationally.
- Integrity—is the ability to have confidence that the data within any system has not been altered and is original as it entered the system.
- Availability—is the ability to access information at any time as and when required.

For many organizations, the business is governed by the requirement to ensure the confidentiality of their data for regulation purposes. However, this is where the main difference exists between IT systems and ICS. The three factors exist in both, but their importance differs and in ICS is referred to as the AIC triad. This change reflects the priority of these types of systems. Availability is the priority factor; the justification for this difference is that ICS requires immediate responses to be made to input data to prevent catastrophic events occurring, meaning that systems and their components need to be available 100% of the time. Integrity in ICS systems is the second priority because the processing in systems is real time which means that they must be able to respond and react to data immediately.

Challenges of OT Security

The problem is that the OT has several peculiarities that make the implementation of the protection measures that are usually adopted for the IT systems difficult and problematic. Systems support the critical infrastructure of the world, and a cyber-attack has the potential impact of loss of life which is more devastating than loss of an IT system. The links between OT and IT have increased during the period of development of the modern world of connectivity. This has increased the vulnerability of systems that could adversely affect communities and the environment.

Availability

The main challenge for OT is based on the availability priority of the AIC triad. OT systems will be operational on a 24-h basis every day of the week and normally 365 days a year. OT systems support the infrastructure of the nation and therefore need to be available. There have been attacks such as the 2000 Maroochy water system (Slay & Miller, 2007), the 2010 Stuxnet attack (Hagerott, 2014), and others that have caused major blackouts and water supply issues which are all effects of non-availability of OT systems. This requirement will lead to systems becoming more vulnerable overtime as they will not have current patches installed, and to apply such maintenance requires advanced complicated plans to ensure there is no disruption to system availability. Another consideration for availability is the real-time nature of these OT systems. The large amounts of data that are generated and analyzed are used instantaneously to alter the system state. The implementation of such security as firewalls and encryption would cause delays in communication and processing which would affect the response and sensitivity of such systems. This could compromise the system operation and ultimately cause loss of life.

There is an additional issue associated with availability, and that is the effects of implementing a patch. It is very difficult with OT systems to test that a patch works before it is implemented on the real system. This inherently increases the risk that a change may influence the operation of another element of the system.

Access Routes

As well as OT systems being at risk from the issues surrounding availability, they also suffer from the challenges of the routes of access provided for such systems. The IoT also have the challenges of access as multiple devices are linked together as a network and could use any one of the following communication media to communicate with other devices:

- Wi-Fi
- Ethernet
- Bluetooth
- Mobile network
- Satellite
- Fiber

The problems that the access route generates are varied and affect different industries who have different requirements. There are some common challenges which are identifiable. One is the ability to send a signal over either a short or long distance. Another is the reliability of the communication medium; if poor weather conditions affect the communication, then that cannot be implemented in an area where this type of weather is common. Other issues could be whether the media is shared by business and residential customers. This could influence the availability of slots to send messages as there could be bottleneck periods such as Christmas and New Year. The speed of communication is very important for ICS systems because of the real-time working environment; some media only offer slow speeds. Another is the potential for interference generated maliciously or unintentionally. Interference can affect all forms of communication and can cause catastrophic effects in ICS systems.

One challenge is the security of the media used to send information. This is an issue for all IT-based systems and is a constant source of development by engineers. It is not possible to make a system 100% secure if it is connected to the outside world. However, the aim of any organization is to provide the securest communication that they can. An area of particular concern for ICS and IoT are the protocols that are used for communication. The communication industry developed technology in an ad hoc manner and suffered from the wide variety of technology. The complexity of communication was due to the high number of different protocols that were available having to communicate with each other. To reduce this complexity, the communication industry formulated a plan to standardize the protocols. The first of these was adopted as a standard in 1984 and was known as the OSI Model. These common protocols are well known, and because of this, they are an area of weakness for any organization. Cybercriminals have been able to research these protocols in detail and have been able to identify flaws that will allow them to gain access to devices using the protocol.

Dependencies

ICS are complex in their nature because of the interlinks that have evolved as technology has been introduced and systems no longer work in isolation. This complexity is described in terms of the interdependency and dependency between components. The main risk is that this complexity has wide-reaching effects when failures occur and can include loss of life.

The key authors in the subject of complexity are Rinaldi et al. (2001) who were the initial presenters of the concept of dependencies and interdependencies. These definitions are frequently referred to and are in use in current literature such as the US Department of Energy (US DOE) report (Argonne National Laboratories, 2015) who quote Rinaldi et al. (2001) to ensure the consistency of the risk and resilience assessment methodology standards:

- Dependency—the reliance or influence of one infrastructure on another through a connection
- Interdependency—the reliance of influence of two infrastructure on each other with a bidirectional connection

Although the US Department of Energy (US DOE) in their report (Argonne National Laboratories, 2015) uses these isolated definitions, they agree with the view of others (European Union Agency for Cybersecurity, 2017; Lauge et al., 2015) that infrastructures cannot be taken into consideration in isolation of the dependencies and interdependencies that exist. The US DOE explanation is based around the interactions between environments. They take the description of a dependency back to the fundamental concept of a control system in having an input that is transformed and then supplies an output which acts as an input to another environment. They further develop the idea into three different types of dependency such as upstream, internal, and downstream.

Complexity

The nature of the size of ICS systems means that the understanding of the systems complexity may not be complete. This could be for various reasons; it is possible that an industry sector is unable to share information, e.g., the nuclear industry, and it is only during a crisis or failure that this crossover of information occurs. Another reason for misunderstanding complexity is that many of these systems have evolved and this evolution has not created a complete set of information on the systems that are in place. It is difficult to have knowledge of every single element in an ICS system which is the fundamental requirement to identify all the interdependencies and dependencies. Another problem is that there are a lot of legacy systems, and having been in place for maybe 50 years, the experience and in-depth knowledge have disappeared as staff have retired.

The complexity of such systems brings with them a higher level of risk. When working in isolation, control systems were protected. Now that they communicate with others, they do not have the same level of protection. Some industries are not able to share information, and this leaves those interconnected at an increased risk. Collaboration is important to be able to deal with complexity of the system of systems effectively.

3.4 Organization Structures

An organization can in the way that it is organized support and reduce the challenges in ICS security. The elements for an organization to consider are:

- Culture and structure
- Financial
- Policies and procedures

The culture of an organization is understood as the group goal and the working relationships. There are different ways to describe the culture of an organization, and the culture will support the leadership and management of the organization. Factors such as empowerment, formality, communication, goal orientation, and bureaucracy will define the culture, but the challenge is to create a working environment that supports the employees and allows them to feel that they can be honest and open about issues. This is important in ICS because a small mistake could be a disaster and employees must be able to flag these as early as possible to reduce the impact. This is known as a no-blame culture.

The physical structure of the organization is a companion of the culture of the organization. It can be rigid or flexible, and many organizations that are rigid are not able to adapt to new situations. In ICS new situations will be a result of the challenges of the working environment, and the organization structure needs to be flexible enough to be able to adapt quickly and continuously improve.

The financial structure of an organization can also be a challenge. Security is an issue that can need addressing in a reactive manner and not proactive. This means that budgets and formal financial processes must be flexible enough for security teams to be able to respond to challenges as early as possible.

The policies and procedures of an organization are important as they support the organization, the leadership, and the employees to undertake their work in a safe manner. ICS organizations must comply with certain regulations and will therefore have fundamental policies such as security in place. The fact that these are in place does not guarantee that they are being used. The challenge for the organization is to not just have these procedures and policies in place but to make sure that they are followed. As stated earlier it is not assured that an awareness of cybersecurity decreases the risk of a cyber-attack. One of the procedures that can support these challenges is the continuous development process. This allows organizations to learn from their experience and improve their processes.

4 Conclusion and Future Work

In this chapter we have reviewed the definitions of the IoT and ICS and compared them to identify the similarities that they have. The chapter has discussed the challenges surrounding these new environments taking into consideration the operation of Industrial Control Systems. The use of ICS too describes the challenges and identifies some of the issues surrounding the operation of real-time systems. The IoT is a system that operates in real time, and therefore the challenges are similar.

In the future this work is to be developed and evolved to not only identify the challenges but also to develop some solutions to these challenges that can be utilized across the residential and commercial environments.

References

Alblabi, S. M., & Weir, G. (2018). User characteristics that influence judgment of social engineering attacks in social networks. *Human Centric Computing and Information Sciences, 8*(5).

Aldawood, H., Alashoor, T., & Skinner, G. (2020). Does awareness of social engineering make employees more secure? *International Journal of Computer Applications, 177*(38), 45–49.

Argonne National Laboratories. (2015). *Analysis of critical infrastructure dependencies and interdependencies*. US Department of Energy.

Assenza, G., & Setola, R. (2019). Operational technology cybersecurity: how vulnerable is our critical infrastructure? *Contemporary Macedonian Defence, 19*(37), 9–20.

Babu, B., Liyas, T., Muneer, P., & Varghese, J. (2017). Security issues in SCADA based industrial control systems. In *2nd International conference on anti-cyber crimes, Abha*.

Bodungen, C. E., Singer, B. L., Shbeeb, A., Hilt, S., & Wilhoit, K. (2017). *Hacking exposed industrial control systems* (1st ed.). McGraw-Hill.

Bostrom, R. P., & Heinen, S. J. (1977). MIS problems and failures: A socio-technical perspective. Part I: The causes. *MIS Quarterly, 1*(3), 17–32.

Centre for the Protection of National Infrastructure. (2021). *Internet of things and industrial control systems*. Centre for the Protection of National Infrastructure [Online]. Retrieved April 1, 2021, from https://www.cpni.gov.uk/internet-things-and-industrial-control-systems.

Collins. (2016). *Collins English dictionary and thesaurus*. HarperCollins.

Doan, D. (2006). *Commercial Off the Shelf (COTS) security issues and approaches*. Naval Postgraduate School.

Ernst and Young Global Limited. (2020). *Global information security survey*. Ernst and Young Ltd.

European Union Agency for Cybersecurity. (2017). *Communication network dependencies for ICS/SCADA Systems*. European Union Agency for Cybersecurity.

S. I. Extraction. (2014). *Project SHINE findings report*. Creative Commons.

Ginter, A. (2016). *SCADA security what's broken and how to fix it* (1st ed.). Calgary.

Greengard, S. (2015). *The internet of things* (1st ed.). MIT Press.

Hadnagy, C. (2011). *Social Engineering The art of human hacking* (1st ed.). Wiley Publishing.

Hagerott, M. (2014). Stuxnet and the vital role of critical infrastructure operators and engineers. *International Journal of Critical Infrastructure Protection, 7*, 244–246.

Hayden, E., Assante, M., & Conway, T. (2014). *An abbreviated history of automation and industrial control systems and cybersecurity*. SANS Institute.

Johnson, S. (2013). Safety and security in SCADA systems must be improved through resilience based risk management. In C. Laing, A. Baddi, & P. Vickers (Eds.), *Securing critical infrastructures and critical control systems: Approaches for threat protection* (pp. 286–300). IGI Global.

Khan, M. A., & Salah, K. (2018). IoT security: Review, blockchain solutions, and open challenges. *Future Generation Computer Systems, 82*, 395–411.

Kilman, D., & Stamp, J. (2005). *Framework for SCADA security policy*. Department of Energy.

Knapp, E. D., & Langill, J. T. (2015). *Industrial network security: Securing critical infrastructure networks for smart grid, scada, and other industrial control systems* (1st ed.). Syngress.

Kriaa, S., Bouissou, M., & Laarouchi, Y. (2019). A new safety and security risk analysis framework for industrial control systems. *Institute of Mechanical Engineers, 233*(2), 151–174.

Kupiers, D., & Fabro, M. (2006). *Control systems cyber security: Defense in depth strategies*. Idaho National Laboratories.

Lauge, A., Hernantes, J., & Sarriegi, J. M. (2015). Critical infrastructure dependencies: A holistic, dynamic and quantitative approach. *International Journal of Critical Infrastructure Protection, 8*, 16–23.

Madakam, S., Ramaswamy, R., & Tripathi, S. (2015). Internet of things: A literature review. *Journal of Computer and Communications, 3*(5), 164–173.

Malatji, M., Von Solms, S., & Marnewick, A. (2019). Socio-technical systems cybersecurity framework. *Information and Computer Security, 27*(2), 233–272.

Maslow, A. (2013). *A theory of human motivation* (1st ed.). Wilder Publications.

Miller, M. (2015). *The internet of things how smart TV's, smart cars, smart homes, and smart cities are changing the world* (1st ed.). Que.

Mouton, F., Leenen, L., & Venter, H. S. (2016). Social engineering attack examples, templates and scenarios. *Computers and Security, 59*, 186–209.

National Institute of Standards and Technology. (2008). *Guide to industrial control systems security*. NIST.

National Institute of Standards and Technology. (2011). *Managing information security risk*. US Department of Commerce.

National Institute of Standards and Technology. (2021, April 15). *COTS*, National Institute of Standards and Technology [Online]. Retrieved April 15, 2021, from https://csrc.nist.gov/glossary/term/commercial_off_the_shelf.

Oosthuizen, R., & Pretorius, L. (2016). Assessing the impact of new technology on complex sociotechnical systems. *South African Journal of Industrial Engineering, 27*(2), 15–29.

Postscapes. (2020, 1 January). *Internet of things (IoT) history* [Online]. Retrieved March 30, 2021 from https://www.postscapes.com/iot-history/.

Rinaldi, S. M., Peerenboom, J. P., & Kelly, T. (2001). Identifying, understanding and analyzing critical infrastructure interdependencies. *IEEE Control Systems Magazine, 21*(6), 11–24.

Simon, T. (2017). *Critical infrastructure and the internet of things*. Centre for International Governance Innovation and Chatham House.

Slay, J., & Miller, M. (2007). Lessons learned from the Maroochy Water Breach. In *International conference on critical infrastructure protection* (Vol. 253, pp. 73–82). Springer.

Techopedia. (2021, April 1). *Legacy system. Janalta Interactive* [Online]. Retrieved April 15, 2021, from https://www.techopedia.com/definition/635/legacy-system.

Trist, E. (1981). The evolution of socio-technical systems a conceptual framework and an action research program. In *Perspectives on organizational design and behaviour* (pp. 19–75). Wiley & Sons.

Tucker, M. (2015). *TE framework: A framework for securing COTs applications*. SANDIA National Laboratories.

United States Nuclear Regulatory Commission. (2016). *Historical review and observations of defense-in-depth*. Brookhaven National Laboratory.

Wu, P. P.-y., Fookes, C., Pitchforth, J., & Mengersen, K. (2015). A framework for model integration and holistic modelling of socio-technical systems. *Decision Support Systems, 71*, 14–27.

Part II
Digital Forensics and Machine Learning in the IoT

An Introduction to Cryptocurrency Investigations

Iqbal Azad ⓘ

1 Cryptocurrencies

1.1 An Introduction

The story of cryptocurrencies began in 2008 with the publication of the bitcoin whitepaper; "Bitcoin: A Peer-to-Peer Electronic Cash System" (Satoshi Nakamoto, 2008) by a person or persons using the name Satoshi Nakamoto. To date that person(s) has not come forward. The whitepaper described a:

> ...system of A purely peer-to-peer version of electronic cash would allow online payments to be sent directly from one party to another without going through a financial institution....

The whitepaper went on to document an electronic payment system based on cryptography, where users would send and receive the "electronic cash" by digitally signing transactions using their credentials, the credentials taking the form of cryptographic public and private keys. The currency was named bitcoin in the whitepaper, and it remains to this date the most widely known and used cryptocurrency.

It was developed as an answer to the centralised nature of finance at the time and the financial crisis in 2008. Bitcoin (shortened to BTC) provided cheaper transaction costs operating on a system which was importantly decentralised. The bitcoin protocols and transactions operate outside of any central government, thereby providing a layer of privacy. There are no physical bitcoins, only balances kept on a public digital ledger that everyone has access to. All bitcoin transactions are verified by massive amounts of computing power.

I. Azad (✉)
London, UK

© The Author(s), under exclusive license to Springer Nature Switzerland AG 2022
R. Montasari et al. (eds.), *Privacy, Security And Forensics in The Internet of Things (IoT)*, https://doi.org/10.1007/978-3-030-91218-5_5

Cryptocurrencies can be argued to be a form of money. Money in its traditional form has several characteristics that make them suitable as a mechanism for the transfer of value. Fiat money is a government-issued currency that is *not* backed by a physical commodity, such as gold or silver, but by the government that issued it. The value of fiat money is derived from the supply and demand, and the stability of the issuing government, rather than the worth of a commodity backing it as is the case for commodity money. Most modern paper currencies are fiat currencies, including the US dollar, sterling and the euro (Merriam-Webster, n.d.).

Cash is a common medium of exchange accepted by many parties as a method for settling economic debt. This can be said to be true of bitcoin and other cryptocurrencies; it is accepted by an ever-growing number of vendors on the Internet and not just the preserve of some criminal marketplaces. Cryptocurrencies are a transferable store of value much like cash; the value of cryptocurrencies can be derived from the effort used to generate them (mining) and the supply and demand. Cryptocurrencies can be transferred from one user to another by creating and "sending" a transaction.

One aspect of cryptocurrencies which has not made it as attractive as a medium of exchange is its volatility in value, due to the fluctuating values against fiat currencies; the value of a product/service may be difficult to predict from one moment to another.

There is a relative inefficiency around cryptocurrency transactions as compared against other payment methods, Bitcoin can currently process between 2 and 4.5 transactions per second (https://www.blockchain.com/charts/transactions-per-second, n.d.); PayPal can process 193 per second; Ripple (XRP) a centralised cryptocurrency, can handle 1500 transactions per second, but the Visa payment system can process 1700 per second.

Cash is recognisable and interchangeable; one £10 note is same as another (save for the serial number); it represents the same value and is recognised globally. Cryptocurrencies as their use has spread is also recognised widely; one bitcoin is worth the same as another bitcoin, although there are differences which will be detailed later.

Cash is divisible to (in the case of sterling GBP) to pennies; as are cryptocurrencies bitcoin has a similar concept, namely, the "Satoshi", which represents 100 millionth of a bitcoin. Therefore, bitcoins can be split into smaller units to ease and facilitate smaller transactions and represented by a number of "Satoshis".

Cash is easy to use because it is transportable (in small amounts); they are conveniently sized and can fit into pockets and wallets. In the digital age, money or traditional currencies don't need to be physically transported. They are accessible via online banking services; transactions can be completed using online transfers or contactless payment methods. Cryptocurrencies do not have a physical manifestation (other than some forms of wallets); they are a purely digital manifestation of value held on individual digital ledgers. Because they exist digitally, they are as transportable as cash, either held online with various services or held on physical devices such as laptops, mobile phones, or USB devices.

Security and the difficulty to counterfeit is another feature of currencies; we wouldn't use a currency if it has no inherent security or was easy to counterfeit. The trust in the currency as a store of value would not be present or justified. Physical cash has security features such as serial numbers, holograms and special construction. Cryptocurrencies such as bitcoin have other security measures such as their decentralised nature and the use of cryptography in generating transactions.

The security of the bitcoin protocol relies on one of its fundamental characteristics, the transaction blockchain. Each transaction since the start of bitcoin in 2008–2009 is published for all to view; copies of all transactions are held on thousands of servers around the world, forming an immutable record of all transactions, which is resistant to alterations. To change any transaction in the past, one would have to locate a change each of those individual records.

However, if a user or company has poor security, it should not be a surprise if their holdings of cryptocurrencies are compromised. Similarly, while banks are relatively secure, it is the customers who are subject to attack by criminals in terms of hacking, phishing and social engineering.

1.2 A Brief History of Bitcoin[1]

August 18, 2008—The domain name is bitcoin.org is registered, current registration details are protected.

October 31, 2008—A person or group using the name Satoshi Nakamoto makes an announcement on the Cryptography Mailing list at metzdowd.com: "I've been working on a new electronic cash system that's fully peer-to-peer, with no trusted third party". This now-famous whitepaper is published on bitcoin.org, entitled "Bitcoin: A Peer-to-Peer Electronic Cash System".

January 3, 2009—The first bitcoin block is mined, block 0. This is also known as the "genesis block" and contains the text: "The Times 03/Jan/2009 Chancellor on brink of second bailout for banks".

January 8, 2009—The first version of the bitcoin software is announced on the Cryptography Mailing List.

January 9, 2009—Block 1 is mined, and bitcoin mining starts.

January 12, 2009—The first BTC transaction is sent between two people and the only other person known to have been sent bitcoin by Satoshi Nakamoto. Satoshi Nakamoto sent 50 BTC to Hal Finney in block 170. The cost of the transaction was 0 BTC (BTC Transaction ID f4184fc596403b9d638783cf57adfe4c75c605f6356fbc91338530e9831e9e16).

---

[1] All the above transactions can be viewed in any number of public or open, "block explorers" for bitcoin. These block explorers are public websites where all transactions are published. They may also provide other functions/services.

October 12, 2009—The first known sale of BTC in exchange for fiat occurred when a Finnish developer Martti Malmi sold 5050 BTC for $5.02, with the dollar amount being transferred via PayPal. The number of BTC sent corresponds with the fact that the only way bitcoin could be obtained back then was by mining it, when the Coinbase reward (reward for mining a block) was set at 50 BTC.

May 22, 2010—Laszlo Hanyecz's 10,000 BTC famous pizza purchase (https://bitcointalk.org/index.php?topic=137.0, n.d.) resides in bitcoin block 57043. This records the purchase of a pizza for 10,000.99 BTC (the 0.99 BTC was to cover the miner's fee) (BTC Transaction ID a1075db55d416d3ca199f55b 6084e2115b9345e16c5cf302fc80e9d5fbf5d48d). The current value of the BTC sent is $560,962,500.00.

June 23, 2011—The Mt. Gox (a now defunct cryptocurrency exchange) CEO Mark Karpeles sent 442,000 BTC from one address to another; many saw this as Karpeles was demonstrating the strength of holdings at Mt. Gox. It remains one of the largest amounts of BTC ever to be sent at one time (BTC Transaction ID 3a1b9e330d32fef1ee42f8e86420d2be978bbe0dc5862f17da9027cf9e11f8c4).

July 1, 2014—30,000 BTC was sent to the winning bidder in an auction held by US Marshals liquidating assets seized from the infamous TOR market Silk Road. The winning bidder was Tim Draper;[2] that purchase, for approximately $18 million, is now valued at $1.6 billion (BTC Transaction ID 9e95c3c3c96f57527cdc649550bf8e92892f7651f718d846033798aee333b0c3).

January 4, 2015—One of the most famous hacks occurred in January 2015 when 20,000 BTC was stolen from Bitstamp; this included a transaction of 3100 BTC which started the theft (BTC Transaction ID a32697f1796b7b87d953637ac827e11b84c6b0f9237cff793f329f877af50aea).

1.3 Cryptocurrency Operations

The operation of cryptocurrencies is determined by its underling protocols in the case of bitcoin by the bitcoin whitepaper.

Transactions are made with an input and output or sender and receiver. The sender and receiver are identified by a bitcoin address. A bitcoin address is the signifier for a store of bitcoins. This is what is recorded on the distributed ledger whenever a transfer of bitcoins is made.

The first ever bitcoin address was generated on the January 4, 2009, when the bitcoin network and system were turned "on". That address was

1A1zP1eP5QGefi2DMPTfTL5SLmv7DivfNa.

---

[2] https://en.wikipedia.org/wiki/Tim_Draper, Tim Draper is famous venture capitalist.

Bitcoin addresses are formatted to a specific structure, namely, 27–34 alphanumeric characters. Each address is unique, and there are

2ˆ160 bitcoin addresses

That number is

1,461,501,637,330,902,918,203,684,832,716,283,019,655,932,542,976

Addresses can be generated by any user of bitcoin. It is also possible to get a bitcoin address using an account at an exchange or online wallet service. One can also generate it offline and store in physical form such as a paper "wallet" or digitally in "wallet" software.

Bitcoin addresses are generated from hashing the cryptographic public key corresponding to a private key generated from the ECDSA or Elliptic Curve Digital Signature Algorithm. Specifically, it uses a particular curve called secp256k1 (https://en.bitcoin.it/wiki/Secp256k1, n.d.). Therefore, in simple terms a bitcoin address is a hash of a cryptographic public key.

Bitcoins and other cryptocurrencies are stored in wallets, but unlike a physical cash wallet, these "wallets" don't store the cryptocurrency themselves. Wallets will contain a public key or address that is used to receive cryptocurrency. It will also contain the private key that is used to verify that you are indeed the owner of the cryptocurrency that you're trying to spend (see Fig. 1). Because of the complexity involved in generating bitcoin addresses, wallet software will compute the public/private key pair for you. The public address is then provided to others to allow them to send you cryptocurrency. It is essential to keep your private keys secure for obvious reasons; losing control of them will mean a loss of control of your cryptocurrency holdings.

This public/private mechanism ensures safety of the cryptocurrency stored but leads to the user having to repeatedly generate a random pair of private/public addresses (or keys) and back them up. As the number of transactions increases, this process becomes cumbersome for the user.

In Fig. 1 we can observe that the public key or bitcoin address 1BgGz* has a private key; this private key can only be used for that corresponding public key or bitcoin address.

Hierarchical deterministic (HD) wallets remove this problem by deriving all the addresses from a single master seed. All HD wallets use a variant of the standard 12-word master seed key. HD wallets eliminate the need for the user to generate and wait for the secure keys to be generated; the users only need to worry about ensuring a backup is created.

Example Private Key: 5HpHagT65TZzG1PH3CSu63k8DbpvD8s5ip4nEB3kEsreAnchuDf

Example Public Key: 1BgGZ9tcN4rm9KBzDn7KprQz87SZ26SAMH

Fig. 1 Image detailing a public/private key pair

The distributed ledger as it names suggests a distributed record of every transaction in a particular cryptocurrency. An entire record of the transactions resides in many servers or **nodes** that help the bitcoin network operate. Not only do they keep a record of all transactions, but they also ensure that transactions are verified. The nodes also operate as a communications network, relaying transactions across the network to all participants. Each of the full nodes separately follows the exact same rules as set out in the bitcoin protocol to decide which blockchain is valid. At this time there are approximately 9600 bitcoin nodes in operation.

Running a bitcoin node entails downloading the bitcoin core software https://bitcoin.org/en/bitcoin-core/ and then running it on a computer. Operating a node means keeping an entire record of all bitcoin transactions and making it available for other users on the network. The record of all transactions is referred to as the **blockchain** and is currently 342 GB in size (as of May 2021).

Blockchains are sequences of individual blocks; each block contains a record of all cryptocurrency transactions completed within a given period. In the bitcoin network, this is approximately every 10 min. Each block has an associated cryptographic hashing problem. This problem must be solved in order for the block to be created and added to the blockchain. Users of the blockchain are rewarded with cryptocurrency for solving these problems; the solution is included in the block and is the "proof of work".

Miners are an essential feature of the cryptocurrency system; they operate as a swarm of ledger keepers. Mining is the process of adding new transaction records to the blockchain. Mining activity is regulated by the bitcoin protocol and is designed to be resource-intensive and difficult so that the number of new blocks of transactions verified each day by miners remains constant. Individual blocks of transactions must contain a proof of work to be considered valid. This proof of work is verified by other bitcoin nodes each time they receive a new block of transactions.

Mining is also the process by which liquidity is added to the bitcoin system. Miners are paid transaction fees as well as a "reward" of newly created coins. These both serve the purpose of supplying new coins and a means of incentivising others to provide security for the system. There will only be a maximum of 21 million bitcoins in circulation; this figure will be reached by 2140, and currently there are 18.699 million bitcoins in circulation. The current reward is 6.25 bitcoins for each block; this is a reduction from the initial reward of 50 bitcoin in 2009. This reward or block reward is determined by the bitcoin protocol. The reward halves every 4 years to regulate the amount of BTC which is newly created when new blocks are mined. When the reward is 0, no new BTC will be created, and the circulating amount will be 21 million.

Because of this relative scarcity, to obtain cryptocurrency, you can either earn it for goods and services, buy it from a cryptocurrency exchange, steal it, or mine it. Each of those methods, other than mining, will have an audit trail of the origin of the cryptocurrency. With mining, the successful miner will earn newly "minted" cryptocurrency.

In the other examples, there will be a record of where you received the cryptocurrencies from, a preceding transaction which can be investigated to identify the source. Likewise, when you spend/send your cryptocurrency, you will create a transaction transferring the value from one cryptocurrency address (yours) to another (recipient). The recipient will then in time send their cryptocurrency on in another transaction and so forth. This will create a series of transactions going all the way back to the origin of the cryptocurrency in the block reward.

1.4   Cryptocurrency Transactions

Understanding how cryptocurrency transactions operate is essential to the investigator. We will use the example of Bob and Alice. Bob wants to send some cryptocurrency say bitcoin to Alice. Bob accesses his wallet software and checks his balance. He has enough to send Alice 1 bitcoin (or BTC as it is denoted), which is what Alice requires as payment. Alice checks her wallet software and, to receive the BTC, takes one of the previously generated BTC addresses. She then communicates the BTC address to Bob. The simplest way to understand cryptocurrencies and bitcoin is to use the example of an old-fashioned accounting ledger.

You will have a personal ledger, which is a personal bitcoin wallet; the credits and debits are recorded in each corresponding column. Your balance is derived as the difference between the credits and debits. One cannot have a negative balance as you cannot send/spend what you do not have. Credits or **received** transactions are made to bitcoin addresses under your control, and the debits or **withdrawals** are then made from the balance of bitcoin that you have.

All your transactions are recorded in your ledger and are also recorded in the distributed ledger or the blockchain.

Going back to Bob, having received the BTC address from Alice, he then accesses his wallet software and creates his transaction sending 1 BTC to Alice. He enters in her receiving address and signifies that he is the owner of the BTC he is sending, by signing the transaction using his private key. In doing so this unlocks the access to his store of BTC.

Using wallet software simplifies the steps required to create a transaction and removes the need to use the CLI (command line interface). The user in this case, Bob, will also enter the fee that he wants to pay the miners for processing the transaction; the higher the fee, the more incentive there exists for the miner to process that transaction. There are a few cryptocurrency wallet types:

* *Software wallets* are programs on digital devices such as phones and laptops. These connect to the network and allow the spending of cryptocurrencies in addition to holding the credentials that prove ownership. Examples would be Electrum, Exodus, Coinbase Wallet and Bitcoin Core. Later versions of wallets will hold multiple types of cryptocurrency.
* Internet services called *online wallets* offer similar functionality; cryptocurrency credentials are stored with the online wallet provider. These are accessed by way of a username and password. Examples would be Coinbase.com and

Blockchain.info. Those online services later also served as cryptocurrency exchanges allowing users to buy and sell cryptocurrencies for cash. Blockchain.info is described as a non-custodial service where the user is in control of the public and private keys, whereas Coinbase.com is a custodial service where the public/private keys are held by Coinbase and therefore custodied or "kept safe".

- *Physical wallets* also exist and are more secure, as they store the credentials necessary to spend bitcoins offline. They can be paper wallets which were popular in the beginning of cryptocurrencies. Nowadays hardware solutions such as Ledger and Trezor wallets are common. These are hardware devices much like a USB stick which store the cryptocurrency credentials.

Once Bob has entered in the requisite details, he via his wallet software sends his transaction message to the bitcoin network. This operation is completed by his wallet software connecting to the nearest and most reliable bitcoin network node. The bitcoin node which receives his transaction message will then verify that the transaction is correct and that you have the necessary funds to send. The node will then tell its neighbouring nodes about the message. Each node will also then tell its neighbours about the transaction messages it has received. In time all the transaction messages will be propagated across the network of nodes.

All the transaction messages which have been received by the network are corralled together in a virtual waiting room. This waiting room is termed the "memory pool" or "mempool". The mining groups or companies will access the mempool and verify a number of transactions. That selection of transactions is influenced by the transaction fee that is being offered by the sender. Importantly each transaction messages comprises a size in data terms; miners will validate and verify 1 MB worth of transactions.

Completing the task allows the miners to be eligible to earn the mining reward; each mining group will attempt to solve a mathematical puzzle, which involves hashing data (viz. all the transaction message data) to achieve a predetermined target value in the resultant hash. This involves computational guesswork and is the reason for the high cost of cryptocurrency mining; the amount of computational power and electricity required to solve the problem is prohibitive to most people.

Hence most cryptocurrency mining is the preserve of large, well-capitalised companies in countries where electricity is cheap. The first miner to come up with a 64-digit hexadecimal number or "hash" that is less than or equal to the target value wins the race to earn the block reward and publish the confirmed block of transactions.

The winning mining group will then communicate its solution or proof of work with the block of confirmed transactions to the bitcoin network. Once this is accepted by the nodes and the proof is checked, they will propagate this across the network to the point that each node will have the latest block mined by miner. This occurs every 10 min; therefore mining can be very lucrative and significant, and investment is put into these mining operations. The current block reward for bitcoin is 6.25 BTC.

Fig. 2 Example of transaction details from Coinbase Analytics TXN ID e9bf2a4f81d086c85 ddd9879769528962719430030899a0e1a8378d1ce8d1f5f

Transaction	
ID	e9bf2a4f81d086c85ddd9879769528962719430030899a0e1a8378d1ce8d1f5f
Time	2017-12-10 01:59:09
Confirmed on	2017-12-10 01:59:09 (block: 498499)

> Amount
> ₿ 6.04724269
> $ 87109
>
> Fee
> ₿ 0.0012
> $ 17

Fig. 3 Example of transaction details from Coinbase Analytics TXN ID e9bf2a4f81d086c85 ddd9879769528962719430030899a0e1a8378d1ce8d1f5f

This block which has been newly mined or confirmed by the miner includes Bob's transaction to Alice. The new block is appended to the previous block of transactions and contains a reference to the block that came immediately before it. Because each block contains a reference to the prior block, the collection of all blocks in existence can be said to form a chain or a blockchain.

Alice can now check her wallet software to see if her BTC have arrived, as her transaction from BOB has been "confirmed" once as being correct by the miners and published in the latest block. In practice Alice will need to wait until the transaction has been in the blockchain for at least three cycles to withdraw funds. When another new block has been appended to the block which contains her transaction, it can be said that her transaction has received **two confirmations**. When another block is added, then her transaction will receive three confirmations and so forth.

Turning to a transaction (Figs. 2 and 3), in this transaction we can see the BTC address 3PLvvrRn1aPoWZcu8GYT2JFCVaF5aqMdm sent 6.04604269 BTC, although the total amount sent is different as this encompasses the fee to be paid to the miner.

Figure 2 displays the recipient address in the transaction as 1PTxj1SCD7Pxk8J k1uoa33xAePJxGh8QvC. The transaction has a unique identifier, namely, transaction hash of e9bf2a4f81d086c85ddd9879769528962719430030899a0e1a8378d1ce 8d1f5f. This transaction appeared in block number 498499 and has 183,566 confirmations since that time.

Fig. 4 Example of change address from Coinbase Analytics TXN ID c6dc0038a9e949359 4a501cc57dafee5d1874d3ae6ae4625571eaae3d39ad0fd

Figure 3 is a simple example of a bitcoin transaction, where the entire amount associated with the input address is sent to the recipient. However, one feature of bitcoin transactions is that the output of any transaction must equal the input (minus the transaction fee). So, in the situation where the output is less than the input, the bitcoin protocol will create change, in the same fashion as a purchase with cash where the amount spent is smaller than the currency handed over to the vendor.

Figure 4 provides an example of a change address. In this case, the wallet software will generate a new bitcoin address and sends the difference back to this address. This is known as the **change**. This change address is controlled by that same wallet and is available to be used as an input for future transactions from that wallet.

In Fig. 4 we can observe this rule, address 3BEXxv* sends 2.301 BTC to 3Egzy9*. A change address is created, namely, 3Avj36*, which has a value of 0.0170, the difference between the input and output (minus fees). The address where the change is sent back to must also be the controlled by the same entity who sent the coins as you would not give your change to another person.

As with change in the fiat world, you will reuse your change for other future transactions combining it with other cash you may have in your wallet or pocket. With bitcoin the concept applies, the newly created change address which has a value associated with it can be used to service a future transaction.

In Fig. 5, the transaction reveals that different types of bitcoin addresses are in use:

- **P2PKH (Pay to Public Key Hash) Legacy Address Format**

    P2PKH is one of the oldest bitcoin addresses in the crypto world and is still a legacy bitcoin address format that is used in the crypto world. It is not segregated witness (segwit) compatible; users can still send bitcoins to other segwit addresses. Transactions with P2PKH are slightly costlier than other segwit addresses because these addresses are longer and take bigger space. P2PKH addresses always start with a 1; 1BvBMSEYstWetqTFn5Au4m4GFg7xJaNVN2 is an example of P2PKH address.

- **P2SH (Pay to Script Hash) Address Format**

    P2SH is newer than P2PKH and starts with 3 unlike 1 in P2PKH. P2SH is slightly complicated than P2PKH and has several functionalities. Transactions with P2SH are more elaborate and have high-security features including a multi-signature facility. This means more than one private key is used to sign transactions. This could infer more than one person or entity in control of a

Fig. 5 Example of reuse of a change address from Coinbase Analytics TXN ID 943503e973ec2b14ee16618094669f6ea1a7eaf0fdefc7a22130a80731b40263, 3Avj36* is combined with other BTC addresses to create a transaction

wallet. The 34-character long address allows multiple digital transactions with multiple addresses and at lower fees compared to P2PKH.

- **Bech32 Segwit Address**

  Bech32 is a new bitcoin address and is the most advanced one compared to the other two addresses. It starts with "bc1" and is longer than P2PKH and P2SH. Bech32 is a segwit address and supports multiple wallets and other addresses. Transactions with Bech32 are faster, and fees are lower.

All three addresses are compatible with each other (they can be sent and received from each other). However, some addresses may have wallet restrictions especially the older bitcoin wallets that may not recognise bech32 segwit addresses. P2PKH or legacy address seems to be the most compatible compared to the other two, but they have higher transaction fees.

2 Attribution and Clustering of Cryptocurrency Transactions

When analysing cryptocurrency transactions and in particular, bitcoin, another phrase used is a cluster or wallet. A cluster or wallet is a group of cryptocurrency addresses which are said to be controlled by a single entity. Thinking back to wallets, they contain the master keys necessary to generate new addresses for receiving or sending cryptocurrencies. Therefore, if we can identify the owner of one address from research, we can associate the other addresses with that owner as well.

2.1 Attribution of Cryptocurrency Transactions

Firstly, taking attribution of addresses, attribution can be obtained via open-source information, research conducted on the Internet. In some cases, this research will reveal ownership of bitcoin addresses. An example of this would be the following video on YouTube, https://www.youtube.com/watch?v=_BoMIxeH8ow. At 12.50 the presenter reveals a bitcoin deposit address for his account at Crypto.com (another cryptocurrency exchange).

09/16/2020

SPECIALLY DESIGNATED NATIONALS LIST UPDATE

The following individuals have been added to OFAC's SDN List:

KARASAVIDI, Dmitrii (Cyrillic: КАРАСАВИДИ, Дмитрий) (a.k.a. KARASAVIDI, Dmitriy), Moscow, Russia; DOB 09 Jul 1985; Email Address 2000@911.af; alt. Email Address dm.karasavi@yandex.ru; Gender Male; Digital Currency Address - XBT 1Q6saNmqKkyFB9mFR68Ck8F7Dp7dTopF2W; alt. Digital Currency Address - XBT 1DDA93oZPn7wte2eR1ABwcFoxUFxkKMwCf; Digital Currency Address - ETH 0xd882cfc20f52f2599d84b8e8d58c7fb62cfe344b; Digital Currency Address - XMR 5be5543ff73456ab9f2d207887e2af87322c651ea1a873c5b25b7ffae456c320; Digital Currency Address - LTC LNwgtMxcKUQ51dw7bQL1yPQjBVZh6QEqsd; Digital Currency Address - ZEC t1g7wowvQ8gn2v8jrU1biyJ26sieNqNsBJy; Digital Currency Address - DASH XnPFsRWTaSgiVauosEwQ6dEitGYXgwznz2; Digital Currency Address - BTG GPwg61XoHqQPNmAucFACuQ5H9sGCDv9TpS; Digital Currency Address - ETC 0xd882cfc20f52f2599d84b8e8d58c7fb62cfe344b; Passport 75 5276391 (Russia) expires 29 Jun 2027 (individual) [CYBER2].

Fig. 6 US Department of Treasury advisory note of new additions to the OFAC (Office of Financial Asset Control) sanction's list

That address is 3HUGYSDXWfdXUH2xKXEBcZpDsnxS5AUZUM. Another example can be seen on the website https://kontestacja.com/, where donations can be made via bitcoin on their sponsors' page: https://kontestacja.com/sponsoruj. On that page a BTC address is displayed with a QR code. That BTC address is 19KvYKUT6hdoJfUrffX36nwo3pkJHtqTXD. These two examples can then be presented as attributions, with the attendant details of the provenance.

In addition to this, there are also circulars and documents regularly posted by government agencies which detail-specific cryptocurrency addresses used in a variety of criminal activities. For example, an advisory note issued by the US Department of Treasury (https://home.treasury.gov/policy-issues/financial-sanctions/recent-actions/20200916, n.d.) lists a Russian individual who is subject to OFAC (US Dept. of Treasury) sanctions (Fig. 6); it also lists the individual's cryptocurrency addresses.

Another example is the US Dept. of Justice indictment (https://www.justice.gov/opa/press-release/file/1304276/download, n.d.) and seizure of BTC addresses linked to terrorism financing of the al-Qassam Brigade (Figs. 7 and 8). One of the bitcoin addresses detailed is 17QAGVpFV4gZ25NQug46e5mBho4uDP6MD.

Figure 6 provides an example of the detail published by government agencies when adding individuals to a sanction's list. The advisory from OFAC lists the individual's name, date of birth and email address. Importantly known cryptocurrency addresses attributed to him are also published.

Figures 7 and 8 detail excerpts from the indictment where a partial bitcoin address is documented and the screenshots are taken from the relevant social media

24. In January 2019, the al-Qassam Brigades began a fundraising campaign on social media to solicit BTC donations from supporters. To receive BTC donations, the organization created multiple cryptocurrency accounts, including one beginning with 17QAW that the organization publicly posted on its social media accounts. The campaign asked donors to send BTC to its accounts, included to 17QAW, as shown below:

Fig. 7 Excerpt from the indictment from the US Dept. of Justice on the financing of Hamas/al-Qassam brigades

Fig. 8 Image from the indictment against al-Qassam Brigades at https://www.justice.gov/opa/press-release/file/1304296/download

for the al-Qassam Brigades. Again, the information published here can be used to inform the attribution of the bitcoin address 17QAW*.

Other sources of attribution can be obtained from direct interactions with services such as cryptocurrency exchanges. In order to obtain this type of attribution, an investigator will need to set up online accounts at those exchanges; to do so one will have to submit legitimate identification to verify your identity.

Most if not all cryptocurrency exchanges have to abide by anti-money laundering rules and financial regulations; their activities are regulated and audited by government agencies to ensure compliance with such laws. A facet of anti-money

Fig. 9 Deposit address for a Binance cryptocurrency account, credited to the author

laundering laws is that financial services should know who their customers are (so-called KYC requirements). This is where the need for identification verification is borne out of; regulated exchanges operate online identity verification systems.

Being regulated means that if an exchange suspects illicit activity by customers, then they are obliged to file a report with the authorities; such reports are referred to as **Suspicious Activity Reports** (SAR) or Suspicious Transaction Reports (STR).

Once you have created an account, a deposit address can be generated, and BTC can be sent to it, thereby creating a transaction and a record of it (Fig. 9). This can be repeated for multiple exchanges and for different cryptocurrencies, thus building up a data set of addresses used and attribution. Withdrawal transactions can also be generated providing for attribution of the sending addresses for an entity.

Figure 9 is a user-generated deposit address for Binance. In similar fashion, addresses for other services can also be generated and monitored. Security researchers and other investigators can use these methods to identify the cryptocurrency addresses for illegal services as well. However, engaging in such activities with illegal marketplaces may make you criminally liable if you have no legal authority to conduct such activity.

2.2 Clustering of Cryptocurrency Transactions

Clustering is the method by which collections of cryptocurrency addresses are associated with a single controlling entity. A cluster can also be described as a wallet. Clustering involves the analysis of the blockchain transactional data to provide this association.

Once an attribution is made for a cryptocurrency address, clustering rules or heuristics will aggregate it with other cryptocurrency addresses according to specified rules. The rules that will operate on a collection of addresses will depend in part, on what pattern of transactions are observed. For example, within bitcoin transactions, all inputs to a transaction must be controlled by a single entity to sign the transaction using a private key (shared spending).

In the case of a transaction with multiple input addresses, it can be deduced that these must be in a single wallet or cluster. Therefore, if one of the addresses used in such a transaction input is known as say Coinbase (from research), then the other inputs must also be controlled by Coinbase (see Fig. 7 for an example).

If a change address from a transaction is identified and then used on a subsequent transaction (with other inputs), then again, all the inputs must belong to a single entity or cluster. Change addresses can be identified in some cases as being the smaller of the output or similar type of address as the input addresses.

The paper (Papagiannaki et al., 2013) "A fistful of bitcoins: characterising payments among men with no name" describes the way in which such heuristics can be used to create clusters of addresses and provide an identity (Fig. 10).

In Fig. 10 an example is shown of change address clustering: in the first transaction, a change address C is identified, and a second transaction is then made which generates a change address F. The third transaction uses two inputs C and F (the change transactions from previously), to send BTC to address G, generating a change address H. Using the shared spending and change address rule/heuristic,

Example

TXN 1
Input: Address A → *Output: Address B*
  Output: Change Address C

TXN 2
Input: Address D → *Output: Address E*
  Output: Change Address F

TXN 3
Input: Addresses C & F → *Output: Address G*
  Output: Change Address H

Cluster Addresses A, D, C, F & H

Fig. 10 Example of clustering methodology using change address analysis

we can conclude that addresses A, D, C, F and H must all be in the same wallet or cluster and under the control of a single entity.

This analysis is dependent upon the correct identification of the change addresses; if from some research we understood that that address D was an exchange, then from the rules, we could conclude that addresses A, D, C, F and H were all also Coinbase controlled. From this example it can be shown that using a combination of attribution data and clustering methods is possible to provide an identity to a collection of addresses.

However, the identity information provided by clustering should be corroborated by other information obtained by the investigator. Conducting this type of analysis takes effort and time, but there are several blockchain analysis tools available; each of them has their own proprietary algorithms or methods for providing attribution and clustering.

2.3   Open-Source Intelligence

Open-Source Intelligence (OSINT) is information obtained from the Internet via research being conducted. It can be defined as:

> the collection, evaluation and analysis of materials from sources available to the public, whether on payment or otherwise to use as intelligence or evidence within investigation (Wells & Gibson, 2017)

There are legal, ethical and moral limits on what should be conducted during an investigation, and those limits will depend on your profession and corporate risk appetite. Poorly conducted OSINT can damage the reputation of the company/organisation, put personal safety at risk (through leaking of information, otherwise known as doxing), jeopardise an ongoing criminal investigation, or taint any potential investigation by the authorities. An important point to consider when conducting OSINT is to ensure that any such activity is proportionate and ethical to the objectives one wishes to achieve.

When embarking on any investigation, the investigator should consider the following:

- What is the purpose of the investigation? What am I seeking to achieve?
- Can I achieve these aims and objectives within the law?
- Do I have the necessary skills, knowledge and equipment to undertake the activity safely?
- To what extent do I need to conduct OSINT? How far should I go and how deep should I dig?
- What are the ethical considerations?
- In conducting this activity, have I assessed the risks involved?
- Are they acceptable and proportionate to the incident or crime I am investigating?
- Where am I going to record my findings and rationale?

Documenting the above (in some form of case management system) will help focus the investigation and ensure that there is no "mission creep". Before embarking on any OSINT, ensure you have considered the above and have the equipment to protect your electronic footprint. When conducting OSINT, it is best to keep a contemporaneous log or record of your actions, to allow for replication of steps and as a record of your activities for review. It may be worthwhile to consider the following points:

- Case summary of incident/crime.
- Aims and objectives/investigation plan.
- Methods to be used.
- Address any ethical/proportionality concerns.
- Any justification for the above points.
- Risk assessment of the methods to be used, e.g. is there a risk of compromise? How have you addressed the risks?
- Record of the OSINT activity with date and time stamp (start/end time, any events of significance).
- Record full URLS in any log of the events.
- Include and/or reference any screenshots taken during the OSINT activity.

The conclusions drawn in an investigation should be based on the information gathered. Always seek corroboration where possible. There is a responsibility to ensure that any conclusions drawn are not solely based on assumptions, suppositions, inaccurate inferences, or falsehoods which could cause unwarranted harm or suspicion. A good rule of thumb is:

Assume nothing, Believe no one & Check everything (**ABC**).

Equipment has been mentioned so it would be wise to review what you will need to conduct OSINT in respect of cryptocurrencies. During your investigation, you may want to conduct transactions. To that end you may want to consider the following:

- Computer (laptop or desktop).
- Secure and stable Internet connection (Use a VPN).
- Hardware wallet.
- Exchange accounts with completed KYC (to purchase and store cryptocurrencies).
- Clustering and attribution tool for cryptocurrencies.
- Screen-capture tools; most modern laptops have this built in.
- A method for making contemporaneous notes.

Address

1FX1TNvp7uPPBtFUSYkTsGbLv5kPziQNni

Summary	Balance

Address		🔖
ID	1FX1TNvp7uPPBtFUSYkTsGbLv5kPziQNni	
Activity	2016-03-13 21:03:40 ↔ 2017-07-21 00:03:06	
Attributes	empty	
Transactions	42	

Wallet	Poloniex ▇	🔖

Fig. 11 Image from Coinbase Analytics, a cryptocurrency analysis tool, detailing the attribution to the address 1FX1TNvp* to Poloniex, a US-based cryptocurrency exchange

2.4 Clustering Tools

Clustering tools are commercially available (at cost); there are a few instances of free tools such as www.walletexplorer.com. This provides a text-based interface for users to review different entities who have been identified and clustered. Users can search by BTC address for any matches to corresponding clusters. Walletexplorer.com also presents data on criminal markets and services such as mixers. Commercial versions have more functionality and attribution data, but there is a cost. One such tool is Coinbase Analytics (Fig. 11).

In Fig. 11, Coinbase Analytics presents the address 1FX1TNvp* as belonging to Poloniex, another exchange based in the USA. Analysis tools will ingest the blockchain data and run heuristics across the information to provide attributions and clusters. One can navigate the data graphically (in some form of mapping) or via a text-based explorer. Various data can be presented such as the balance of an address or cluster, as well as the totals in and out of an address/cluster.

- **Exchanges**

    Cryptocurrency exchanges or Virtual Asset Service Providers (VASP) exist to provide the on/off ramp into cryptocurrencies. Customers can create online accounts with such a service and then buy or sell cryptocurrencies with fiat money. There are hundreds of such exchanges worldwide catering to different geographic locations, customer needs and cryptocurrencies. Some also provide professional trading platforms where advanced and complex investment strategies can be used to speculate on the performance of different cryptocurrencies. The most popular exchanges can transact in billions of dollars a day.

    Most cryptocurrency exchanges operate within a system of financial regulation. These regulations vary according to different countries, but generally this

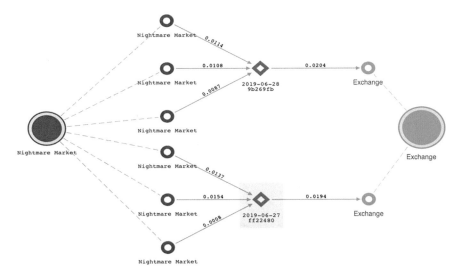

Fig. 12 Coinbase Analytics graph showing the movement of BTC from Nightmare Market to an exchange. The diamond shapes indicate a transaction, hollow circles are addresses and larger ones are clusters of addresses

means that exchanges must know who their customers are, prevent criminal activity and report suspicions to the authorities.

This is important to note because even with the popularity of cryptocurrency currently, people still need to convert cryptocurrencies to fiat cash to realise any gains. In the context of investigations, this is the main point of weakness for criminals. When investigating the flow of funds from a criminal service, you are looking for the nexus with a regulated exchange, who will have records on file of the customer in question (Fig. 12 details the flows of funds from an illicit marketplace to an exchange).

In Fig. 12, BTC from Nightmare Market is observed to flow into an exchange account. It is important that when following funds, you do not continue to follow the funds through an exchange. This is because they deal with the deposit of cryptocurrencies much in same way a bank will deal with a deposit of cash. When you deposit say a £10 note at a bank, it will not be the same £10 that you withdraw from the ATM outside the bank later. Banks operate floats of cash, having enough to cover withdrawals from the bank on any given day. When it runs low, they will top it up with reserves.

Cryptocurrency exchanges operate in the same manner, maintaining a float or a hot wallet (hot wallets are connected to the wider Internet, and constantly used and therefore potentially vulnerable. Cold wallets are offline storage of cryptocurrencies). This hot wallet will contain enough cryptocurrency to meet customer withdrawals on a day; it can be topped up. Conversely if there is too much, the excess can be sent to secure storage or a cold wallet (offline storage of

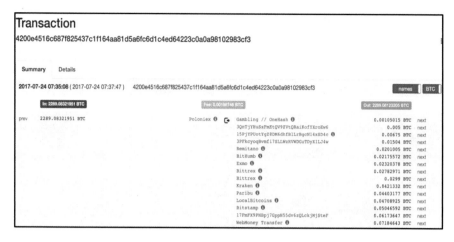

Fig. 13 Transaction from Coinbase Analytics detailing a withdrawal from Poloniex

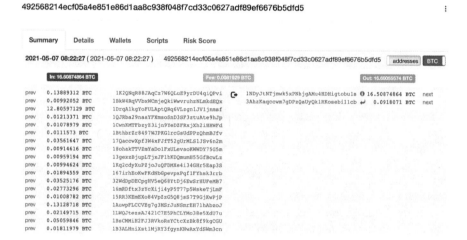

Fig. 14 Coinbase Analytics—example of Binance consolidating deposits into one large UXTO

cryptocurrency). The amount of cryptocurrency held in the hot wallet will depend on a company's risk tolerance.

Customers will deposit into this hot wallet, which will then be used to service withdrawals for other customers. Exchanges will re-use deposits; this breaks the chain of transactions and makes investigation very difficult without the customer information an exchange can provide (Fig. 13).

In Fig. 13, one large Poloniex input is used to send to many different destinations; this is an example of what is termed a "batched transaction". Exchanges will consolidate many different inputs or deposits from customers

Transaction

2bc60d2316f6a5a72f3e0a1d1b1cb6a64bc9e6d5a1a40e034327abf954e8291c

Fig. 15 Coinbase Analytics—Example of Binance UXTO being used to service withdrawals. In this example there are two inputs as each relates to a separate transaction into that address. It is then represented as separate UXTOs

| prev | 4.01805632 BTC | 1NDyJtNTjmwk5xPNhjgAMu4HDHigtobu1s ⓘ |
| prev | 30.98993128 BTC | 1NDyJtNTjmwk5xPNhjgAMu4HDHigtobu1s ⓘ |

Fig. 16 Coinbase Analytics—separate UXTO input in Fig. 15 transaction

into one large input BTC address. This address or unspent transaction output (UXTO) is then used to send to many addresses; this reduces the fees involved in processing many individual withdrawals in single transactions (Figs. 14, 15 and 16).

In Fig. 14, the transaction has many inputs from what we know are Binance deposit addresses. Binance is a large well-established cryptocurrency exchange. The inputs are then sent or consolidated to the bitcoin address beginning 1NDyJ*. This address is well known as a Binance address.

In Fig. 15, which is a subsequent outbound transaction from Binance, the inputs are again 1NDyJ*. This is the exchange using the consolidated inputs from Fig. 14, to service many customers' withdrawals. In this example there are two inputs as each relates to a separate transaction **into** that address. Each output in the transaction represents an individual customer's withdrawal or send.

In Fig. 16, we can see the separate inputs for that transaction; each represents a separate deposit into the address, and they are represented as separate UXTOs for the address 1NDyJtN*. As we have observed in the transaction, both inputs combined have sufficient funds to service all the customers' withdrawals.

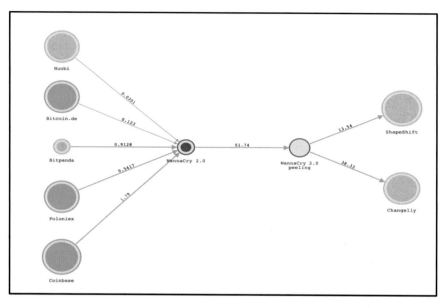

Fig. 17 Coinbase Analytics graph of the outputs from the WannaCry ransomware attack in 2017

- **Coinswapping Services**

    Coinswapping services are a non-custodial service; they do not store or hold your cryptocurrencies but offer the functionality to exchange one coin for another. They aggregate the best prices for a particular cryptocurrency pair and allow you to then swap the pair. An example of a coinswapping service is Shapeshift.com.

    For example, if I wanted to exchange BTC for another coin say Ethereum, I would send BTC to the service and stipulate the receiving wallet address for the Ethereum. Once the BTC transaction has been confirmed, the equivalent amount of Ethereum is then sent to my address.

    All of this is completed without taking the customer deposits; this is achieved by using another exchange as the provider of the cryptocurrency. The coinswapping service acts as the middleman in the transaction. These services will again break the chain of transactions making it difficult to follow the funds (Fig. 17).

    In Fig. 17, the graph depicts bitcoin ransoms being paid by victims of the ransomware. These originated from known and regulated exchanges. The BTC was moved from the WannaCry 2.0 wallet to a secondary wallet named WannaCry peeling 2.0. From this wallet, the funds were then sent to coinswapping services.

    We can review one of the transactions from WannaCry peeling 2.0 to Shapeshift (TXN ID f7866ba8bea329e800ad71b71dac21acc6fc9f996c82690f2f 9776eb71664841).

    This transaction is a swap from BTC to another currency; Shapeshift operates a query to identify the currency that was "swapped" into. Enter the URL https://

{"status":"complete",
"address": "16Edzk6CqjZVyvGUAHFeAznM4Da396ua2R",
"withdraw":
"46jECnrkJTUks7Fg5YtwShdUCiwUwEEZJtJBhKLK4GBWfBX7PLrY-
BuWR9zhzos5uQ1uXGUgFpGCSBR5o651pL5ERxmwqHu",
"incomingCoin":1.8161984,"
incomingType":
"BTC","outgoingCoin":"109.8593376",
"outgoingType":
"XMR","transaction":"a6ebd8cbba75153786e8d4ff471df1003e0d047afe84f1eba0
56bd42b1f7afb5",
"transactionURL":"https://xmr-
chain.net/tx/a6ebd8cbba75153786e8d4ff471df1003e0d047afe84f1eba056bd42b1f
7afb5"}

Fig. 18 Text output from query https://classic.shapeshift.com/txstat/16Edzk6CqjZ VyvGUAH-FeAznM4Da396ua2R

classic.shapeshift.com/txstat/. Then paste the corresponding deposit address, at the end of the URL. In the case of the WannaCry Ransomware case, the deposit address was 16Edzk6CqjZVyvGUAHFeAznM4Da396ua2R.

The output from this query is then provided:

In Fig. 18, the query results are returned to the operator as text output. The nature of the Shapeshift swap transaction is then revealed; it was a swap of 1.8161984 BTC for 109.8593376 XMR or Monero. Monero is another cryptocurrency which is unfortunately completely untraceable. This query is specific to Shapeshift; other coinswapping services such as ChangeNOW and Changelly do not have such a function.

- **Darknet Markets**

    Darknet markets are TOR (The Onion Router)-based marketplaces where illicit goods and services are advertised for sale. It is akin to eBay for criminals, with vendors advertising products/services for sale. Many of the products are illegal in their nature, hence the use of TOR. The method of payments in most cases, is via cryptocurrencies, initially bitcoin but as the marketplaces have evolved other cryptocurrencies has been adopted.

    AlphaBay, the largest criminal darknet market, was taken down by authorities in 2017; it was used by hundreds of thousands of people to buy and sell illegal drugs, stolen and fraudulent identification documents and access devices, counterfeit goods, malware and other computer hacking tools. The site operated as a hidden service on the TOR network to conceal the locations of its underlying servers as well as the identities of its administrators, moderators and users.

The alleged administrator, Alexandre Cazes, committed suicide in a Thai prison shortly after being apprehended.[3]

AlphaBay was not the first nor was it the last darknet market, since Alphabay's take down other market sites that have sprung up each having their own unique selling point, https://darknetlive.com, and other similar sites maintain a list of darknet markets and the associated URLS/domains.

Each darknet market operates differently; however many operate an escrow system whereby customers will deposit cryptocurrencies into a deposit address. This deposit addresses are usually the customers' own individual address, when purchases are made; the relevant funds are moved to an escrow wallet and then only moved to the seller when delivery of the product has been verified. The seller will then withdraw their cryptocurrency or balance should they wish. The operators of the market site will take a fee from every transaction; a business model is very much like that of eBay. Some marketplaces and vendors also operate "tip jars" or donation addresses; a well-known market was that of Dream Market. That address was

1DREAMv7k16T8bMyE7ghe4nLQVydBbPJAe

Having identified a cluster of address used by a darknet market, we can observe the flow of funds into and out of it. Investigations on darknet markets usually concentrate on the outbound flows of funds to identify those behind the marketplace. The product being sold will inform the focus of any cryptocurrency analysis in terms of values being sent and received by that market.

In Fig. 19 we can see BTC moving from Nightmare Market to Shapeshift, Bittrex and Binance. Clusters of addresses are indicated by solid circles. This view is at the wallet/cluster level and not transactional. The movement of funds are shown as arrows with the relevant total amount and direction. Legal applications can then be made for customer data from the different exchanges, if appropriate.

The cluster- or wallet-level analysis will reveal the relationship between two different wallets/clusters and will represent the total amount of crypto sent and received between them.

- **Mixers/Tumblers**

    Cryptocurrency mixers or tumblers operate to obfuscate the flow of cryptocurrencies; they attempt to sever the links from the origin of the funds to the intended recipient. They accomplish this task by mixing the transactions with other likeminded users.

    Bitcoin and other cryptocurrencies were developed in part to provide greater financial privacy. Many users of cryptocurrencies are also privacy conscious in

---

[3] https://www.dailymail.co.uk/news/article-4714656/Justice-Dept-announces-takedown-online-drug-marketplace.html

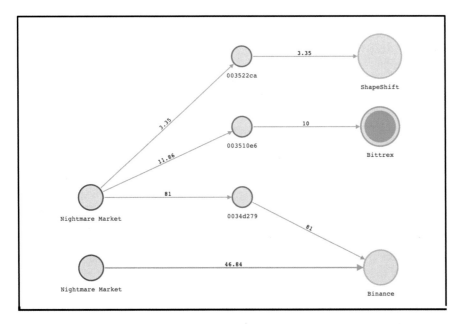

Fig. 19 Coinbase Analytics graph of funds moving from Nightmare Market

that regard but using cryptocurrencies does not provide you with total privacy but is a pseudo-anonymous method of conducting financial activity.

The blockchain is a fully auditable with an entire record of all transactions. With the correct tools, the transactions can be traced to an exchange or service where the end user can be identified. Because of this transparent record, cryptocurrency mixing or tumbling services have sprung up as a method to provide more privacy. Such services have been used by criminal elements to launder the proceeds of exchange hacks, darknet markets, fraud and other illicit activities. There are many mixing methods that have been utilised over time, from fully centralised solutions where users trust a single mixing service to decentralised versions where no trust is required (other than the underlying mixing protocol).

Mixers will ingest the transactions from several users and then use these inputs and create a number of different onward transactions in varying amounts, co-mingling the various inputs in many transactions. This severs the connections between the inputs and outputs. At the end of the process, the intended recipients will receive the cryptocurrency. Many centralised mixers also have a reserve of cryptocurrencies to allow for speedier transactions; they do not need to wait for the number of customer deposits to be a certain amount before initiating the mix-ing transaction. Mixing services can be visualised as a blender; cryptocurrencies with a taint or criminal history can be placed in the blender. The blender will then (like a food processor) will break up the ingredients, mixing them all together (including clean reserve coins) before outputting cryptocurrencies with little or no taint.

Centralised mixers are services that accept the payments and send different coins in return. If many users utilise a particular mixing service, it does become difficult for an observer to trace the "incoming" coins to any of the destination addresses. Centralised mixers however require the users to trust the service; the mixer knows exactly which user sent and received which coins, and anyone in control of the mixer's data would have a good starting point to identify who sent what, where and when. If this user data fell into the hands of criminals or seized by police, then the users find themselves being targeted by criminals or (maybe worse) face arrest. Lastly, users trust the service to conduct the transactions they want; should the service refuse to, then there is little recourse for the user.

Decentralised mixers purport to solve these issues; CoinJoin (CoinJoin is a specific mixing protocol) mixers, by letting a large group of users cooperate in making one large payment to their respective recipients. Basically, if a hundred users all send exactly 0.1 BTC to a new address they control, and then merge these 100 transactions into one big transaction, everyone gets 0.1 bitcoin back, but no one can see where they got it from. CoinJoin mixers can be configured to ensure that not even the entity that "merges" the transaction can figure out which coins went where. The protocol cannot steal any coins: Users wouldn't sign the merged transaction if they would not get their 0.1 BTC back.

Helix was a mixing service, which provided a mixing or tumbling service that helped customers conceal the source of funds. It operated by taking a fee for services and was in operation for a 3-year period. Helix allegedly transferred over 350,000 bitcoin, with a value at the time of transmission of over USD $300 million. The operator specifically advertised the service to conceal transactions on the darknet from law enforcement. In February 2020, criminal charges including ML conspiracy and operating an unlicensed money-transmitting business were brought against the operator, Larry Harmon. The US regulator (FinCEN) also laid a civil penalty against Mr. Harmon to the value of $60 million. It is alleged that Helix partnered with the darknet marketplace AlphaBay until AlphaBay's seizure by law enforcement in 2017.[4]

Much like darknet markets, the clusters associated with mixing services can also be identified and mapped out. In the case of the 2019 Binance exchange hack, the stolen BTC was traced to wallets which split the funds from the initial first wallet. The persons behind the theft then used at least two different mixing services, thereby giving them the best opportunities to move funds and not leave themselves with a single point of failure (Fig. 20).

In Fig. 20, the graph details the flow of bitcoin from the theft at Binance to intermediary wallets to mixing services. The flow of funds can be traced through wallets that each took a proportion of the stolen coins to the mixing services.

---

[4] https://www.coindesk.com/fincen-fines-bitcoin-mixing-ceo-60m-in-landmark-crackdown-on-helix-coin-ninja

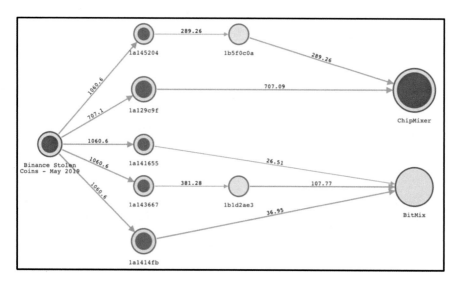

Fig. 20 Coinbase Analytics graph of the 2019 Binance theft

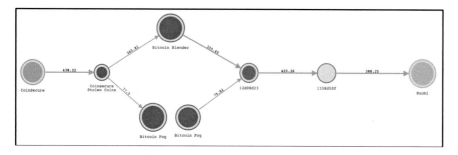

Fig. 21 Coinbase Analytics graph (wallet level) of the Coinsecure theft 2018

In April 2018 Coinsecure (an Indian cryptocurrency exchange) suffered a theft of approximately 438 bitcoins. This was allegedly perpetrated by one of its senior staff (https://www.zdnet.com/article/coinsecure-not-so-secure-millions-in-cryptocurrency-stolen-cso-branded-as-thief/, n.d.). The addresses were publicised and tracked (https://news.bitcoin.com/coinsecure-announces-repayment-plan-bounty-stolen-bitcoins/). Again, we can see that the persons behind the theft have used different mixing services to launder the stolen funds. The perpetrators used a single wallet to receive the coins from both mixing services; from this wallet the stolen BTC can be traced to the exchange Huobi (Fig. 21).

In Fig. 21, investigators followed the stolen bitcoin from Coinsecure to a wallet. This wallet then sent the stolen bitcoin to mixing services. Although the graph in Fig. 21 appears simple, it belies the amount of effort taken to identify the common wallet used to receive the funds from the mixing services. Reviewing each of the identified outputs from each service in a specific period could provide this answer.

Fig. 22 Value of the cryptocurrency market, information taken from https://coinmarketcap.com

Date	Total USD Value (Billions)
May 2013	1.329
May 2014	6.336
May 2015	3.880
May 2016	8.611
May 2017	38.201
May 2018	467.88
May 2019	186.853
May 2020	246.402
May 2021	**2450.814**

When the same wallet is being sent to from both mixing services and the balance received is commensurate with the amounts inputted to the mixing services, then this should arouse suspicion and a determination that the wallet contains proceeds of the Coinsecure theft.

3 The Cryptocurrency Economy

It is important to understand the cryptocurrency economy as it may present opportunities or challenges to the investigator. Bitcoin came into being in 2009; in 2013, the market capitalisation of the cryptocurrency market was $1.329 billion dollars since that time the volumes of cryptocurrency transacted and traded have grown immensely from a niche interest to a trillion-dollar industry. Hundreds of billions of dollars' worth of cryptocurrencies are bought and sold each day on hundreds of exchanges (Fig. 22).

In Fig. 22, the chart details the exponential growth of the cryptocurrency economy in recent years, from a total value of $1.329 billion in 2013 to over $2 trillion in 2021. In 2013 there were a handful of cryptocurrencies available; now there are thousands available on hundreds of different exchanges worldwide which exist as online businesses and services. There are many data aggregation sites which can provide an insight into the values being traded/transacted within cryptocurrency. www.coingecko.com presents lots of different figures for the user. It does provide a useful starting point to explore the ecosystem. CoinGecko lists 7163 different cryptocurrencies in circulation currently (May 2021).

We cannot explore each of them, but a useful starting point is to look at the top 10 coins by value or market capitalisation and review a few of them (Fig. 23).

In Fig. 23, taken from the Coinbase website, the top 10 cryptocurrencies ranked by its market capitalisation are listed. Within the list are a wide variety of different cryptocurrencies. BTC is the most dominant cryptocurrency in circulation accounting for approx. 40% of the total value in the whole cryptocurrency economy.

In the second place is **Ethereum**. Ethereum is not a type of digital currency but a type of decentralised global computing platform; many uses currently deployed on

Name	Price	Change	Price chart	Volume (24h)	Market cap $\updownarrow$	Supply
ⓑ Bitcoin BTC	£33,397.05	-10.21%		£65.9B	£627.8B	18.8M
◈ Ethereum ETH	£2,417.53	-11.57%		£41.1B	£283.9B	117.4M
◈ Ethereum 2 ETH2	£2,417.53	-11.57%		£41.1B	£283.9B	117.4M
⊛ Cardano ADA	£1.68	-13.22%		£11.2B	£54.0B	32.0B
ⓣ Tether USDT	£0.73	+0.29%		£166.8B	£49.9B	68.6B
◉ Binance Coin BNB	£291.72	-13.91%		£4.5B	£49.1B	168.1M
✗ XRP XRP	£0.78	-16.68%		£11.8B	£36.1B	46.5B
≋ Solana SOL	£115.16	-14.53%		£14.4B	£33.9B	292.8M
Ⓓ Dogecoin DOGE	£0.18	-15.37%		£5.7B	£23.2B	131.2B
Ⓢ USD Coin USDC	£0.73			£6.5B	£20.6B	28.4B

Fig. 23 Top 10 cryptocurrencies as of May 2021, taken from Coinbase.com

the Ethereum blockchain include financial tools and games to complex databases. *Mastering Ethereum: Building Smart Contracts and dApps* (Antonopoulos & Wood, 2018) provides an in-depth explanation of Ethereum.

Ethereum-based applications are built using "smart contracts". Smart contracts, like paper contracts, establish the terms of an arrangement between two parties. But unlike an old-fashioned contract, smart contracts automatically complete when the conditions are met without the need for either party to know who is on the other side of the contract and without the need for any intermediary.

Current Ethereum-based applications include stable coins (like DAI, which has its value pegged to the dollar by smart contracts), decentralised finance apps (collectively known as DeFi), and other decentralised apps (or dApps). You interact with the Ethereum network by using ETH to pay a network fee to execute smart contracts and to send ETH to others. As a result, the fees paid in ETH are called "gas".

With bitcoin, addresses signify a store of bitcoin; in Ethereum these are called *accounts*, and there are two types:

- Accounts that only store ETH—these are very similar to bitcoin addresses and are sometimes known as externally owned accounts (EOAs). One can send and receive to these addresses by making transactions.
- Accounts that store ETH and have smart contracts; the smart contracts are activated by a transaction sending ETH into it. Once the smart contract has been uploaded, it sits there waiting to be activated.

0xc00e94cb662c3520282e6f5717214004a7f26888

Fig. 24 An example of an Ethereum address; this one is the smart contract for Compound, an ERC-20 token

In Fig. 24, the Compound cryptocurrency smart contract address is displayed. There are clear differences in formatting and composition between an Ethereum address and a bitcoin address.

There are many differences between bitcoin and the Ethereum system; one difference is that digital assets can be built on top of the Ethereum blockchain, and these assets are called tokens. Developers do not have to create a new blockchain; they can instead use the existing infrastructure of the Ethereum infrastructure. To create a token, the developers must conform to the Ethereum token standards. Tokens can be described as a form of smart contract on the Ethereum blockchain. Some of the most popular tokens are ERC-20, which is a standard for interchangeable tokens, such as Compound (COMP).

ERC-721 is a standard for non-interchangeable tokens, such as piece of digital art. In recent times there has been an increase in the use of ERC-721 or non-fungible tokens (NFTs) to create new markets for art and other collectibles, on the blockchain. Some notable NFT marketplaces are OpenSea and Rarible. Much like traditional artworks, NFTs have a cost and can be used to transfer value between persons or represent an investment. The transfer such as NFTs can be traced on the Ethereum blockchain. In March 2021, Christies sold the NFT artwork *Everydays: The First 5000 Days*, an artist known as "Beeple"; the price was $69 million.[5]

The complexity and utility of Ethereum have lent itself to the creation of what is termed "decentralised finance" or DeFi. As the name suggests, this is a decentralised network of various protocols and applications that allow a user to borrow, lend, buy and sell different cryptocurrencies. The values involved have reached $86.19 billion as of May 2021.

Tether is a type of cryptocurrency which is pegged to that of the real-word fiat currency, in the case of Tether, the US dollar. It was introduced in 2015 (https://tether.to/wp-content/uploads/2016/06/TetherWhitePaper.pdf, n.d.) following the publication of its protocols/whitepaper; it was envisaged to be a cryptocurrency backed by a fiat currency, and that is its main premise. Each Tether coin or dollar is backed 1:1 by $1 in real world. Originally deployed on the bitcoin blockchain using a layer called "Omni Protocol", the use of Tether has migrated to other blockchains. Most of the supply of Tether is now on the Ethereum blockchain, because of its stability; in comparison to other cryptocurrencies, it has become a useful asset for investors and users.

38Hm5dXXUEa6v9WpAziQYSuL2ndr3Q85ga

---

[5] https://www.theverge.com/2021/3/11/22325054/beeple-christies-nft-sale-cost-everydays-69-million

The above is a Tether address on the Omni layer on the bitcoin blockchain. In the next transaction, it can be observed that the Tether moves to Tether address 1ByoveJ8QSG7hxu4JY86tXeuthzZn7juNx. Tether addresses operate in a similar fashion to bitcoin; clustering and attribution information can be ported over. The attribution will be the same for both chains as they operate on the bitcoin blockchain.

TXN ID c1e27b0bbb54eb259b1963c5c6e3c9ceeb4c53f33098344766047f7827 8a48d3

Sender: 38Hm5dXXUEa6v9WpAziQYSuL2ndr3Q85ga

Receiver: **1ByoveJ8QSG7hxu4JY86tXeuthzZn7juNx**

Address 1ByoveJ8QSG7hxu4JY86tXeuthzZn7juNx is attributed to Huobi on the Tether *and* bitcoin blockchain. This type of cross-chain attribution can be used in investigations to provide attribution data, where coins share or have shared a common blockchain.

Bitcoin Cash was created in August 2017 from what is known as a hard fork of the bitcoin blockchain. So-called soft forks also occur within cryptocurrencies; these are less dramatic changes to the underlying protocol, such that the wallets and nodes on that chain upgrade to take advantage of the new rules.

During 2017 disagreements raged between different parties within the bitcoin community over how to improve the efficiency of the bitcoin blockchain. One group comprising some developers and mining organisations launched a different version of bitcoin. The differences in the new protocol meant that the bitcoin blockchain split into two supported chains. At that time if you held 1 BTC on the original blockchain, you would be credited with 1 Bitcoin Cash on the new chain, and your original bitcoin address was valid on both chains.

Bitcoin Cash (BCH) was thus created as the rules or protocols underpinning it were so different. It continues to be in use today and features in the top 10 coins by value. It is not only bitcoin which has been forked. Bitcoin Cash continued to be forked into many different variants from Bitcoin Gold to Bitcoin Diamond, each with a different take on the implementation of bitcoin.

BCH Address format : qphund3xjf2a3ttr653xm3pyfx8kduxleyjw5yvygu

Legacy Address : 1BC5eTS1M9nPnVGt7GQKCVaUFmntaMuVuf

The above are two examples of BCH addresses; the first is the native BCH address called "CashAddr". The format was introduced to remove user error when sending and receiving Bitcoin Cash. The format changed helped differentiate between Bitcoin Cash (BCH) and bitcoin (BTC) addresses. Each BCH address also has a "legacy address", which is its corresponding bitcoin address. Legacy addresses are not normally used due to the similarities with BTC addresses. Cross-chain attribution can be used in the BCH blockchain and bitcoin blockchain.

4   Conclusion

The use of cryptocurrencies is expanding globally, it is a recognised asset class (https://www.cnbc.com/2021/02/08/tesla-buys-1point5-billion-in-bitcoin.html, n.d.); social media stories of life-changing wealth have led to increased interest. That interest has also attracted criminal elements who are all too skilled at exploitation. Initial Coin Offerings (ICO) from 2014 to 2019 were characterised by scams and Ponzi Schemes (Sapkota et al., 2020). The prevalence of such scams has continued today, with the DeFi ecosystem particularly affected by exit scams, price manipulation and fraud. Cryptocurrency use is being observed in most crimes, and investigators will need understand and adapt to this trend. Understanding the concepts of tracing cryptocurrencies will provide tangible lines of enquiry to allow for the attribution of individuals and recovery of assets. Courts in the UK have recognised that cryptocurrencies can be recognised as property albeit intangible (Ion Science Limited and Duncan Johns v Persons Unknown, Binance Holdings Limited and Payment Ventures Inc, 2020) and therefore subject to seizure orders. Clustering and the attribution of cryptocurrency addresses will improve, but challenges do still exist particularly in the use of different cryptocurrencies and the global nature of the trade.

References

Antonopoulos, A. M., & Wood, G. (2018). *Mastering Ethereum building smart contracts and DApps*. O'Reilly Media. ISBN 9781491971918, 1491971916.
https://bitcointalk.org/index.php?topic=137.0 (n.d.).
https://en.bitcoin.it/wiki/Secp256k1 (n.d.).
https://home.treasury.gov/policy-issues/financial-sanctions/recent-actions/20200916 (n.d.).
https://tether.to/wp-content/uploads/2016/06/TetherWhitePaper.pdf (n.d.).
https://www.blockchain.com/charts/transactions-per-second (n.d.).
https://www.cnbc.com/2021/02/08/tesla-buys-1point5-billion-in-bitcoin.html,   https://
    www.cnbc.com/2021/02/24/microstrategy-buys-more-than-1-billion-worth-of-bitcoin-adding-
    to-massive-holdings.html (n.d.).
https://www.justice.gov/opa/press-release/file/1304276/download (n.d.).
https://www.zdnet.com/article/coinsecure-not-so-secure-millions-in-cryptocurrency-stolen-cso-
    branded-as-thief/ (n.d.).
Ion Science Limited and Duncan Johns v Persons Unknown, Binance Holdings Limited and
    Payment Ventures Inc. (unreported, 21 December 2020), & AA versus Persons Unknown
    [2019] EWHC 3556 (Comm).
Merriam-Webster. *Fiat money*. Retrieved from https://www.merriam-webster.com/dictionary/
    fiat%20money
Satoshi Nakamoto. (2008). *Bitcoin: A peer-to-peer electronic cash system*. Retrieved from
    www.bitcoin.org
Papagiannaki, K., Gummadi, K., & Partridge, C. (2013). *IMC '13: Proceedings of the 2013
    Conference on Internet Measurement Conference* (pp. 127–140). https://doi.org/10.1145/
    2504730.2504747

Sapkota, N., Grobys, K., & Dufitinema, J. (2020). How much are we willing to lose in cyberspace? On the tail risk of scam in the market for initial coin offerings. *SSRN Electronic Journal*. https://doi.org/10.2139/ssrn.3732747

Wells, D., & Gibson, H. (2017). OSINT from a UK perspective: Considerations from the law enforcement and military domains. In *Proceedings Estonian Academy of Security Sciences, 16: From Research to Security Union* (pp. 84–113). Estonian Academy of Security Sciences. SSN: 2236-6006 (online).

The Application of Machine Learning Algorithms in Classification of Malicious Websites

Tabassom Sedighi, Reza Montasari, and Amin Hosseinian-Far

1 Introduction

In today's society, our inescapable reliance on technology makes it almost impossible to ignore the ever-present dangers of malicious online resources and the threat that they pose to financial, personal and business security. The computer security company Kaspersky states in their 2016 statistics report that 31.9% of their computer customers were 'subjected to at least one Malware-class web attack over the year' and that '261,774,932 unique URLs were recognised as malicious by web antivirus components' (Garnaeva et al., 2016). These statistics show how important it is to be cautious when using the web. In this chapter, the competency of machine learning algorithms is compared and evaluated with a view to determining how effective these are to detect malicious websites.

These comparison and evaluation are based only on the data that can be obtained from HTTP headers, WHOIS data and DNS records. The advantage of using only this data is that it can all be obtained without the need to parse any code located on the client or the server which could have potentially harmful effects. To achieve the stated objectives, first, the dataset will be heavily pre-processed into an appropriate format so that it can efficiently be utilised. Next, the dataset will

T. Sedighi
Vision and Eye Research Institute, Faculty of Health, Education, Medicine and Social Care, School of Medicine, Anglia Ruskin University, Cambridge, UK
e-mail: tabassom.sedighi@aru.ac.uk; https://aru.ac.uk/

R. Montasari
Hillary Rodham Clinton School of Law, Swansea University, Swansea, UK
e-mail: Reza.Montasari@Swansea.ac.uk; http://www.swansea.ac.uk

A. Hosseinian-Far (✉)
Department of Business Systems and Operations, University of Northampton, Northampton, UK
e-mail: amin.hosseinianfar@northampton.ac.uk; https://www.northampton.ac.uk

© The Author(s), under exclusive license to Springer Nature Switzerland AG 2022
R. Montasari et al. (eds.), *Privacy, Security And Forensics in The Internet of Things (IoT)*, https://doi.org/10.1007/978-3-030-91218-5_6

be subject to sampling, scaling and dimensionality reduction before three machine learning algorithms are applied with the aim of successfully identifying whether the data are describing a malicious or benign website.

2 The Dataset

The dataset selected for this study is 'Malicious and Benign Websites', provided by Urcuqui (2018) on Kaggle. The dataset contains information about 1781 unique websites. Out of these websites, 1565 are benign and 216 are malicious. For each website, the dataset contains 21 attributes of metadata that describe information about the application and network layers of the website, all of which are freely available for public access. The first attribute is named URL, which have all of its values replaced with unique identifying values in order to protect the anonymity of the data. This attribute is therefore relatively futile to a machine learning algorithm as every value is unique and unrelated and will therefore not be used. The last attribute is the 'Type' of the website, i.e. malicious or benign, and is a binary value of 0 or 1 with 0 being benign. Therefore, there are 19 potentially useful attributes in the dataset which can be refined later using dimensionality reduction techniques.

3 Data Preparation

This section provides an outline of how the raw dataset was modified and prepared so that it would be ready to feed into machine learning algorithms. This includes processes such as missing value handling, dimensionality reduction and data normalisation.

3.1 Data Analysis

Of the 19 attributes in the dataset, excluding the URL and Type, 13 contain numerical data, 4 are categorical and 2 contain date-time values. Tables 1, 2 and 3 provide information concerning each of these attributes.

Table 1 Information about numerical attributes

Name	Min	Max	Mean
URL_LENGTH	16	249	56
NUMBER_SPEC IAL_CHARACTERS	5	43	11
CONTENT_LENGTH	0	649263	11726
TCP_CONVERS ATION_EXCHANGE	0	1194	16
DIST_REMOTE_TCP_PORT	0	708	5
REMOTE_IPS	0	17	3
APP_BYTES	0	2362906	2982
SOURCE_APP_PACKETS	0	1198	18
REMOTE_APP_PACKETS	0	1284	18
SOURCE_APP_BYTES	0	2060012	15892
REMOTE_APP_BYTES	0	2362906	3155
APP_PACKETS	0	1198	18
DNS_QUERY_TIMES	0	20	2.26

Table 2 Information about categorical attributes

Name	Unique count	None count
CHARSET	9	7
SERVER	240	175
WHOIS_COUNTRY	49	306
WHOIS_STATEPRO	182	362

Table 3 Information about timestamp attributes

Name	Unique count
WHOIS_REGDATE	DD/MM/YYYYHH:MM
WHOIS_UPDATED_DATE	DD/MM/YYYYHH:MM

3.2 Data Formatting and Conversion

Numerical Attributes

The numerical data attributes were naturally in the correct format to be used by a machine learning algorithm with the only issue being any values that were set to N/A. These were all replaced with the value -1 as there were no other negative values in the dataset, and this allowed for a clear distinction between true values and missing values.

Categorical Attributes

The first decision that was made about the categorical attributes was that the WHOIS_STATEPRO attribute would not be used as it has a very large number of missing values. As a result, it would be unlikely to be useful to the machine learning algorithms. The server attribute also has a high number of unique values and 'None' values. However, the server attribute would almost certainly be an effective indicator if there were enough data entries, and adequate pre-processing was performed on it. Therefore, it was decided to keep this attribute.

The categorical attributes required much more pre-processing before they could be ready to be used by a Machine Learning (ML) algorithm. First, the datasets were analysed, and any values that represented the same category were combined into one category. For instance, given that, in the WHOIS_COUNTRY column, there were values 'us' and 'US', these were converted so that all values referring to the United States would be 'US'. This was applied to all categorical data. For each categorical attribute, the full set of unique values was then indexed, and each occurrence of each attribute was replaced with its corresponding index value. All missing values and none values were then set to -1 for the same reason as with the continuous data.

Timestamp Attributes

The third and final datatype in the dataset is timestamp data; these required a few steps of processing before they could be used. First, any data values that were not in a timestamp format were either converted manually to the correct format or were converted to 'NaT' meaning Not a Time, if the value did not represent date and time information. Then, the date-time values were all converted into integers that represented the time in seconds so that they were in pure integer form, and, finally, all NaT values were converted to be equal to a value that would simulate the -1 that has been used for the other two data formats; the function for this is displayed below:

$$NaT = Date_{min} - Date_{range},$$

where:

- NaT = Not a Time values
- $Date_{min}$ = Lowest time value in the attribute
- $Date_{range}$ = The range of date values

3.3 Random Under Sampling

The dataset being used in this project has a heavy majority of one class over the other. As a result, there are many more benign websites than there are malicious websites. This imbalance can cause overfitting of ML algorithms if it is not dealt with effectively. Since the imbalance is so large, the most appropriate way to address this was to use random under-sampling. Random under-sampling involves reducing the size of that dataset by removing entries from the larger class until the classes contain the same number of instances. This leaves the dataset much smaller than it was originally, but, in some cases, it can significantly improve the accuracy of the ML algorithms, usually on the initially smaller class. Whilst there was the option to employ some forms of oversampling, this would have left the dataset with many copies of every instance in the minority class and could potentially cause bad overfitting to this class.

3.4 Scaling

Once the data have been processed, it could then be utilised for ML; however, the significant variation in data ranges and values can cause the ML algorithms to apply imbalanced importance to the attributes. This issue can be addressed by converting the dataset, so that all attributes have similar statistical attributes such as range, standard deviation or mean. For this study, min-max scaling was selected for a range of -1 to 1. This was due to the fact that the missing data had all been set to equal -1 and that this form of scaling would retain the distinction that was desired when this decision was made. The following provides the functions for min-max scaling:

$$X_{std} = \frac{X - X_{min}}{X_{max} - X_{min}}$$

$$X_{scaled} = (X_{std} * (max - min)) + min$$

- $X =$ Value to be scaled
- $X_{min} =$ Minimum of attribute values

One of the main disadvantages of min-max scaling as opposed to standardisation is that the resulting standard deviations are smaller, denoting that outliers are less easily detected. This has not created an issue for the dataset used in this study since the data are all exact and no errors were made during the collection and production of the data.

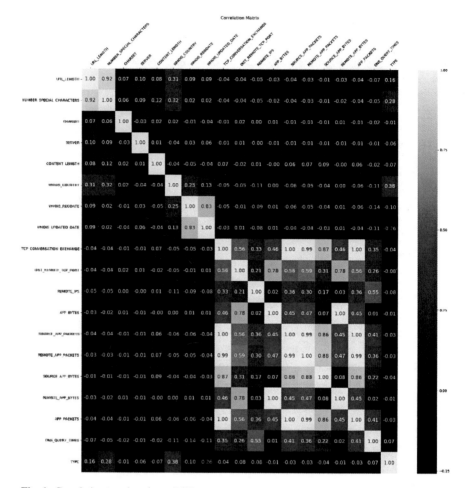

Fig. 1 Correlation matrix prior to PCA

3.5 Dimensionality Reduction

Having a large quantity of data is extremely important to ensure the accuracy and quality of ML algorithms. However, this amount of data requires substantial processing power to be able to perform the required calculations. In order to alleviate this issue, there exist several techniques that can be performed to reduce the amount of data without losing the usefulness that it provides. For this study, principal component analysis (PCA) technique was selected. Before the dimensionality reduction was performed, it was essential to visualise the relationships between the attributes in the dataset. For this purpose, a correlation matrix was generated and plotted as shown in Fig. 1.

As represented by Fig. 1, a large percentage of the attributes in the bottom left of the plot are very closely correlated to one another. However, they show very little relationship to the classification. On the contrary, *URL_LENGTH, NUMBER_SPECIAL_CHARACTERS, WHOIS_COUNTRY* and *WHOIS_UPDATED_DATE* are all strongly correlated with the classification. Nevertheless, they have low correlation with each other, indicating that these features were most likely to be useful to the ML algorithms.

Principal Component Analysis

Having plotted the correlation matrix revealed that some forms of dimensionality reduction could have a large impact on the performance of the dataset. PCA is a dimensionality reduction technique that generates a decreased number of features whilst retaining the highest percentage of the underlying information as possible. In order to achieve this, it combines them by projecting all the original data into lower dimensional space in a manner that makes them no longer realistically interpretable by a human. PCA was then performed in a loop for every number of output attributes up to the original count. Next, the classification algorithms (discussed in the next section) were used to analyse the effectiveness of the PCA. The results of this analysis are depicted in Figs. 2, 3 and 4.

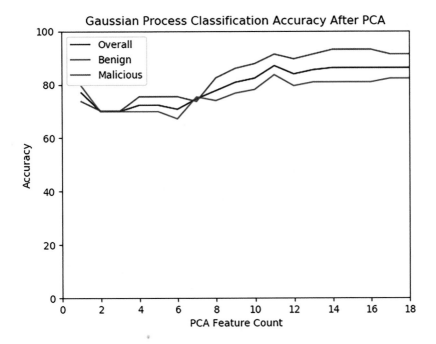

Fig. 2 PCA analysis using Gaussian process classification

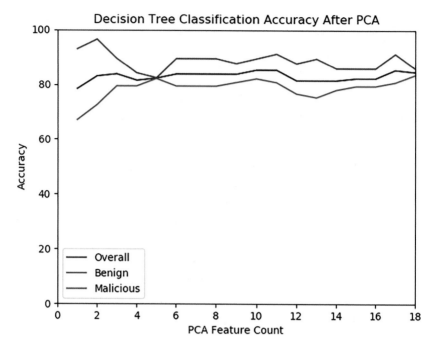

Fig. 3 PCA using decision rree classification

Figures 2, 3 and 4 clearly demonstrate the relevance and usefulness of the PCA to be applied to this dataset as the results for all three classifiers do not show any significant improvement above 11 features, i.e. less than 2/3 of the initial amount of data. In view of this, it was decided to use the feature count of 11 through PCA as this appeared to be the point of plateau/convergence of the Gaussian process classification and the support vector classification. Had the decision tree classification was the sole focus of this study, a much lower feature count could have been selected since the reduction of features appears to have much less effect on this classifier. Once the PCA had been completed with a feature count of 11, a new correlation matrix as displayed in Fig. 5 was produced to show the relationships between the new features.

As the new correlation matrix reveals, the PCA has produced a set of features that have very little correlation with each other. However, these features are almost all correlated to the classification, illustrating that it has successfully achieved its intended purpose.

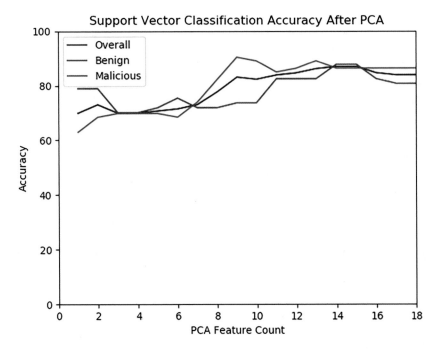

Fig. 4 PCA using support vector classification

4 Classification Methods

As previously stated, this study involved applying three ML classification algorithms to the dataset with the aim of achieving significant prediction accuracy of the classification. The remainder of this section provides an in-depth discussion of these classification methods.

4.1 Gaussian Process Classification (Binary)

The first classification algorithm implemented in this study was Gaussian process classification. This ML algorithm uses a regression model to fit the data and then calculates a probability for each class using this. It then determines the most effective probability to use as a splitting point between the prediction classes to find its output predictions. In this case, the regression model is Laplace approximation. This algorithm is different when dealing with multiple output classes (Williams & Rasmussen, 2006; Daneshkhah et al., 2017; Batsch et al., 2019).

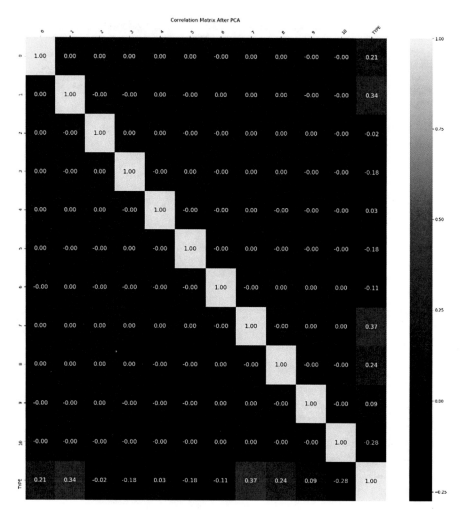

Fig. 5 The new correlation matrix after the completion of the PCA

4.2 Decision Tree Classification

Decision tree classification is an ML algorithm which progressively splits the dataset by incrementally adding rules that provide the largest increase in prediction accuracy. This process terminates when the accuracy is no longer increasing (Grąbczewski, 2014).

4.3   Support Vector Machine Classification

The support vector classifier attempts optimally to separate classes by constructing hyperplanes that split them. These hyperplanes use linear boundaries if possible but can become much more complex when dealing with non-linearly separable output classes (Friedman et al., 2001).

5   Results

Having carried out all the experiments, the results were tabulated and plotted. This section first provides the overall results in Tables 4 and 5, where:

- True negative represents values that were correctly predicted as benign.
- False positive refers to values that were falsely predicted as malicious.
- False negative outlines values that were falsely predicted as benign.
- True positive defines values that were correctly predicted as malicious.

The remainder of this section offers the results concerning the individual classification methods.

Table 4 Data showing the overall accuracy

Overall results	Accuracy (%)	Benign accuracy (%)	Malicious accuracy (%)
Gaussian	86.92	83.56	91.23
Gaussian no PCA	87.69	86.30	89.47
Decision tree	84.62	79.45	91.23
Decision tree no PCA	89.23	84.93	94.74
Support vector	83.85	84.93	82.46
Support vector no PCA	82.31	89.04	73.68

Table 5 Data representing overall result counts

	True negative	False positive	False negative	True positive
Gaussian	61	12	5	52
Gaussian no PCA	63	10	6	51
Decision tree	58	15	5	52
Decision tree no PCA	62	11	3	54
Support vector	62	11	10	47
Support vector no PCA	65	8	15	47

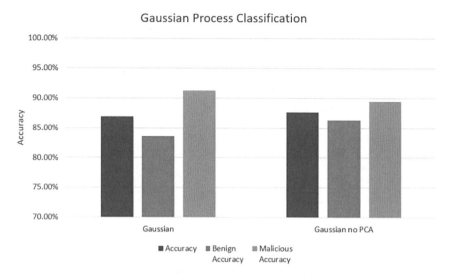

Fig. 6 Gaussian process classification results

5.1 Gaussian Process Classification Results

The fact that the Gaussian process classifier is technically a regression model with a classification layer on top means that the probability of each prediction can be determined as displayed as in Fig. 6. The results represented by Fig. 6 reveals the probability that each website is benign with the truly benign websites being in green and the truly malicious applications being in red. Furthermore, the results demonstrate that whilst the classifier incorrectly predicted around 13% of websites, only a few had a probability that was significantly bad. Most of the incorrect predictions were very close to the cut-off value of approximately 0.4, denoting that, with some fine-tuning, this model could potentially be brought much closer to 100% accuracy. Additionally, Fig. 7 outlines the probability of benign.

Tables 6 and 7 outline the Gaussian process approximations as part of two confusion matrices. Table 6 incorporates principal component analysis (PCA) technique, whereas Table 7 presents results without PCA.

5.2 Decision Tree Classification Results

Similarly, Fig. 8 presents classification made using decision tree (DT), whilst Tables 8 and 9 provide approximations made with decision tree with and without PCA, respectively.

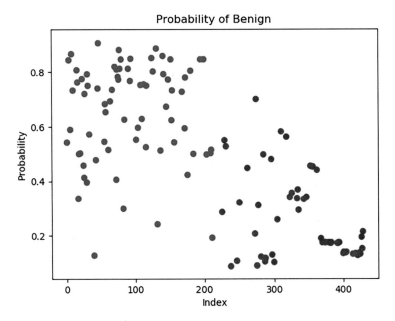

Fig. 7 Gaussian process benign probability

Table 6 Gaussian process confusion matrix (PCA)

Gaussian			Predicted	
			Benign	Malicious
Actual	Benign		61	12
	Malicious		5	52

5.3 Support Vector Classification Results

Results for support vector (SV) classification are presented in Fig. 9. Likewise, the confusion matrices with PCA and without PCA using support vector are outlined in Tables 10 and 11, respectively.

Table 7 Gaussian process confusion matrix (no PCA)

Gaussian no PCA		Predicted	
		Benign	**Malicious**
Actual	**Benign**	63	10
	Malicious	6	51

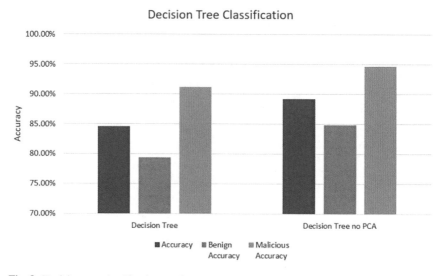

Fig. 8 Decision tree classification results

Table 8 Decision tree confusion matrix (PCA)

Decision tree		Predicted	
		Benign	**Malicious**
Actual	**Benign**	58	15
	Malicious	5	52

Table 9 Decision tree confusion matrix (no PCA)

Decision tree no PCA		Predicted	
		Benign	Malicious
Actual	Benign	62	11
	Malicious	3	54

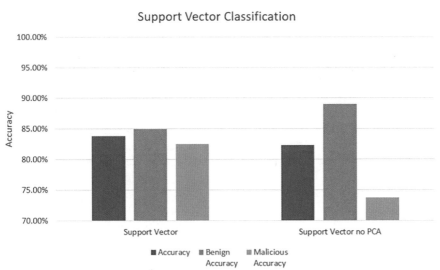

Fig. 9 Support vector classification results

Table 10 Support vector confusion matrix (PCA)

Support vector		Predicted	
		Benign	Malicious
Actual	Benign	62	11
	Malicious	10	47

6   Discussion and Conclusion

The analysis of the results obtained reveal that all three of the ML techniques used in this study have achieved their intended purpose to predict the nature of a website

Table 11 Decision tree confusion matrix (no PCA)

Support vector No PCA		Predicted	
		Benign	**Malicious**
Actual	**Benign**	65	8
	Malicious	15	47

from the provided data. All of the overall accuracies are between 80% and 90% with similar values for each of the classes alone indicating that there is no overfitting of one class. The application of the principal component analysis showed an overall minor reduction of accuracy. However, for the improvement of efficiency, this trade-off is almost certainly worthwhile. The nature of the application incurs a higher cost to the misclassification of malicious websites than it causes to the benign. As a result, the ML technique that would best apply to this context would be one that achieves a higher accuracy on malicious websites than it applies to the benign ones. Furthermore, the results reveal that the Gaussian process classifier or the decision tree classifier would fit that role, whereas the support vector classifier would not be appropriate for the stated role.

Considering the correlation matrix in Fig. 5, it could be deduced that forward selection might have been a better choice of dimensionality reduction due to the high number of attributes with very limited usefulness or potentially even a combination of forward selection and PCA. Therefore, this could be investigated as a future work.

In summary, ML algorithms offer many opportunities to detect malicious websites without the need for high-risk website content parsing. Instead, as the study has shown, this can be achieved by using data from HTTP headers, WHOIS lookups and DNS records. Of the three classifiers used in this study, the Gaussian process classifier is the most appropriate option for the application. This is due to the fact that it is a good balance between effectively managing dimensionally reduced data and achieving a high accuracy on specifically the malicious websites. Another point of consideration for future work could be a further investigation into the possibility of forward feature selection alongside additional other similar ML methods.

References

Batsch, F., Daneshkhah, A., Cheah, M., Kanarachos, S., & Baxendale, A. (2019). Performance boundary identification for the evaluation of automated vehicles using Gaussian process classification. In *2019 IEEE Intelligent Transportation Systems Conference (ITSC)* (pp. 419–424). IEEE.

Daneshkhah, A., Hosseinian-Far, A., & Chatrabgoun, O. (2017). Sustainable maintenance strategy under uncertainty in the lifetime distribution of deteriorating assets. In *Strategic engineering for cloud computing and big data analytics* (pp. 29–50). Springer.

Friedman, J., Hastie, T., & Tibshirani, R. (2001). *The elements of statistical learning* (Vol. 1, No. 10). Springer Series in Statistics.

Garnaeva, M., Sinitsyn, F., Namestnikov, Y., Makrushin, D., & Liskin, A. (2016). Overall statistics for 2016: Kaspersky security bulletin. *Blue book.*

Grąbczewski, K. (2014). *Meta-learning in decision tree induction* (Vol. 1). Springer International Publishing.

Urcuqui, C. (2018). *Malicious and benign websites: Classify by application and network features (dataset).* Kaggle. Retrieved from https://www.kaggle.com/xwolf12/malicious-and-benign-websites

Williams, C. K., & Rasmussen, C. E. (2006). *Gaussian processes for machine learning* (Vol. 2, No. 3). MIT Press.

Part III
Cyber Terrorism and Violent Extremism in the IoT

The Use of the Internet and the Internet of Things in Modern Terrorism and Violent Extremism

Aime Sullivan and Reza Montasari

1 Introduction

In the twenty-first century, the unprecedented rise of the Internet has transformed society and modern communication. Subsequently, as the Internet becomes embedded within our everyday lives, it plays an increasing role in the activities of violent extremists. Extremist individuals have exploited the Internet as a tool to advance recruitment, propaganda, training and communication. According to recent research, 61% of UK terrorists are engaging in online activity directly related to their radicalisation and/or terrorist activity (Gill et al., 2017, p. 107). Therefore, it is critical to understand the factors that accelerate or facilitate violent extremism as these harmful ideologies may empower an individual to take action. Technological advancements such as the Internet of Things (IoT) and cloud computing (CC) have provided extremists with a new global domain for social interaction to bring attention to their specific cause. Nonetheless, contemporary and historical literature has debated the interplay between violent extremism and the Internet. Traditionally there exist two polarised assumptions. On the one hand, it is suggested that the Internet plays a more important role than face-to-face interaction in the violent radicalisation process (Sageman, 2008b). On the other hand, it is argued that the role of the Internet in the process of radicalisation is minimal and consequently overstated (Laqueur, 1999; Burke, 2011). Recently, academic literature has begun to move beyond this dichotomous way of thinking; however the extent to which the Internet enables radicalisation to occur remains a complex and contested issue.

Therefore, this chapter will examine the role of the Internet and associated technology such as the Internet of Things (IoT) and cloud computing (CC) in the

A. Sullivan · R. Montasari (✉)
Hillary Rodham Clinton School of Law, Swansea University, Swansea, UK
e-mail: Reza.Montasari@Swansea.ac.uk; http://www.swansea.ac.uk

process of radicalisation to terrorism and violent extremism. Particular emphasis will be placed upon the role of online environments in facilitating communication and the spread of extremist ideology, as opposed to operational and strategic recruiting functions of the Internet. Before assessing the debate, this chapter shall define two key concepts, radicalisation and violent extremism. Firstly, the chapter will provide an assessment of academic literature that has perpetuated a false offline/online dichotomy. Then, the chapter will demonstrate the new communicative role of technology within violent extremist movements since the emergence of Web 2.0. Following this, Sageman's (2008a) 'bunch of guys' bottom-up framework will be employed to understand how the process of radicalisation can be facilitated and amplified by the digital age, referring to the concepts of moral outrage, othering, virtual communities and echo chambers. Lastly, contemporary literature that considers the interplay between online and offline domains will be explored. Overall, with reference to a growing body of literature, this chapter argues that the offline and online worlds of extremist ideology are considerably intertwined, whereby ultimately the Internet plays a substantial role in radicalisation to violent extremism.

The remainder of this chapter is structured as follow. Section 2 briefly discusses debates relating to the terminology in the realm of terrorism and extremist studies. Section 3 describes the evolution of the Internet and academic debate, whilst Sect. 4 explores pathways to violent extremism in the digital age. Section 5 investigates radicalisation and identity negotiation. Section 6 offers a detailed review of the current debates about the extent to which the Internet and associated technology play a role in radicalisation. Section 7 discusses the convergence of the offline and online worlds of violent extremists. Section 8 examines the role of the Internet of Things and associated technologies in terrorism. Finally, the chapter is concluded in Sect. 9.

2 Definitions

Terminology in the realm of terrorism and extremist studies is widely debated. Firstly, there remains a lack of conceptual clarity regarding the definition of radicalisation. This chapter adopts a broad definition of radicalisation as 'the process by which a person comes to support terrorism and extremist ideologies associated with terrorist groups' (HM Government, 2015, p. 21). This definition focusses upon the emergence of extreme beliefs and values. It is critical to recognise that radicalisation is not an inherently violent phenomenon, and not all radicalisation processes are negative. Only a small minority of radicalised individuals incite violent ideologies or commit acts of terrorism. However, for the purpose of this chapter, radicalisation will be referring to the harmful radicalisation of violent extremists. Secondly, violent extremism is a complex concept that lacks a precise and universally agreed definition. This chapter will utilise the United Nations' (2016, p. 143) definition of violent extremist as 'someone who promotes, supports, facilitates or commits acts of violence to achieve ideological, religious, political goals or social change'. To reiterate, not every individual who possess extremist beliefs will engage in violent acts. Therefore, it is important to define these ambiguous terms in order to differentiate between non-violent extremists and violent extremists and ideologies and action.

3 The Evolution of the Internet and Academic Debate

There remains an ongoing debate as to the role and influence of the Internet in the process of violent radicalisation. Traditionally, academics have been broadly separated into two distinct perspectives: those who believe the Internet has transcended offline radicalisation and those who are sceptical that the Internet has such a significant impact upon the radicalisation process. It is important to note that scholars have now began to acknowledge radicalisation beyond this offline and online dichotomy; nonetheless binary thinking is highly pervasive. There are several scholars who question the ability of the Internet to radicalise individuals. For instance, Walter Laqueur (1999, p. 262), a prominent figure in terrorist studies, remained unconvinced. He argues that the power of information and communication technology does not yet transform into 'real power'. This indicates that the virtual world will have limited effects upon terrorism and violent extremism. Laqueur (1999) arguably contradicts this point by suggesting that audiocassettes played a fundamental role disseminating ideologies in previous extremist movements, thereby acknowledging that extremists can exploit technological developments to spread propaganda. Nonetheless scepticism regarding the prevalence of the online sphere continues in contemporary analysis. Burke (2011) explored an extremist jihadi group and pertains that social media will never replace radicalisation at a grassroots level as face-to-face interaction is a vital component. Whilst Burke (2011) observes that the Internet can aid and facilitate communication, propaganda, recruitment and donations, he does not deem these components as useful as 'real-world' operations.

To counter this scepticism, researchers began to direct analysis towards the understanding of the Internet as a substitute for 'real world' radicalisation. Marc Sageman (2008b) became the strongest advocate for this shift in thinking. Sageman (2008b) is a key scholar that asserts the Internet has enabled the creation of strong social bonds and exposure to extremist materials subsequently transforming jihadist radicalisation. He pertains that the Internet has become a centralised hub of communication whereby individuals can now autonomously support and internalise extremist discourse, thereby providing the opportunity for 'self-radicalisation'. Sageman (2008b) extends this further and argues social interaction that accelerates extremist radicalisation is no longer taking place face to face, but rather online. However, this remains an unsubstantiated claim as although the Internet facilitates communication and radicalisation, scholars should remain cautious as to not provide generalisations about the causality of the Internet's radicalising potential.

Academic scholarship has largely deviated from simplistic analysis of the Internet and radicalisation. However, the exaggerated threat of the Internet persists in political discourse and policy. For example, the government's Online Harms White Paper establishes a duty of care for Internet companies (Whittaker, 2021). However, Whittaker (2021) argues this paper focuses almost exclusively upon harmful content and the online radicalisation of impressionable individuals. Consequently, the interplay between the online and offline environment is overlooked. This perpetuates the narrative that the threat of violent extremism is exclusively caused by the Internet. Consequently, the aforementioned scholars and policy makers indicate that the 'real world' and 'virtual world' are largely distinct realms. This provides a simplistic view of radicalisation and seeks to divide two radicalising environments into dichotomous categories. The Internet has become increasingly integrated into people's everyday lives. The online and offline practices of everyday life are highly dependent upon one another and intersect in very significant ways (Whittaker, 2021). However, critic Burke (2016) began to drastically shift his stance within a later journal article 'How Evolving Media Technology is Changing Terrorism'. He asserts that technological advancement has had a considerable real-world impact upon extremist individuals and organisations. Burke (2016) claims that the Internet is an integral instrument that enables extremists to rapidly disseminate ideological material. This development demonstrates that online radicalisation research simultaneously evolves alongside the rapidly changing nature of technology.

The revolution of the digital world profoundly shaped society and the extremist landscape. As demonstrated, political discourse and academic literature surrounding extremism and radicalisation are dominated by jihadism. ISIS has been valorised as an elite group of Internet users (Conway, 2017). However online radicalisation is not exclusive to jihadi groups or religious ideologies. Thus, it is imperative to diversify research beyond radical Islamist ideologies and consider the wider online ecosystem that incorporates numerous groups, domains, platforms and ideologies. Violent extremist use of the Internet is not a new phenomenon; communication technologies have always been an important instrument in the dissemination of political, social and religious ideologies (Conway et al., 2019). Consequently, in order to better understand the role of the Internet in radicalisation to violent extremism, concepts must be placed within a broader historical context.

The extreme alt-right movement remains closely linked with the emergence of radical Internet and meme culture. Despite this, the far right has historically been overlooked by researchers and policy makers. Radical right supporters began to use the Internet as early as the 1980s to spread white supremacist ideologies and radicalise audiences. Extremists initially utilised computerised bulletin boards such as Aryan Nation Liberty Net to communicate, later followed by online forums (Conway et al., 2019). In the 2000s the ecology of the Internet later shifted to 'Web 2.0'. This era is characterised by an advancement in social media platforms and interactivity. During this technological evolution, extremist discourse and narratives shifted to social media and alt-tech platforms enabling easier propaganda dissemination and two-way communication (Conway et al., 2019). This demonstrates that as the Internet evolves, so does extremists use of it. Thus in the digital era, there are increasing amounts of extreme violent content circulating online. According to Benson (2014), this is not a surprising development. Individuals and extremists will interact through the dominant communication vehicle of that era: forums, radio, print-based media, television, etc. Currently, the Internet is one of the most predominant means of communication. Thus, the increased role of social media in extremist ideological transmission is the result of technological evolution as opposed to a strategic and deliberate choice made by violent extremists. Nonetheless there are fundamental aspects of the Internet that distinguishes it from other communication vehicles and subsequently lends itself to online radicalisation.

4   Pathways to Violent Extremism in the Digital Age

Radicalisation is not a unidirectional procedure; there is no universal pattern or blueprint that explains how people become violent actors. However, that being said, there are a number of factors that are common to most models and theories of radicalisation. This chapter will employ Marc Sageman's (2008a) bottom-up approach to explore the role of social media and the Internet in radicalisation to violent extremism. Sageman (2008a) outlines a four-stage radicalisation process. The first stage involves a sense of moral outrage, for example, an emotive reaction to a specific event or set of grievances. Secondly, this discontent is interpreted as a 'war' and moral violation that evokes an 'us' versus 'them' mentality. Thirdly, everyday experiences of discrimination, bias and unemployment perpetuate feelings of moral outrage. Lastly, establishing a social network enables the exchange of extremist ideologies and content which in turn amplifies existing grievances and network mobilisation and can intensify radicalisation. According to Sageman (2008a), the pathway of radicalisation to violent extremism is non-linear in that there is no sequential progression through the stages. Furthermore, the 'prongs' of this pathway are not mutually exclusive; they can exist in tandem with one another. Radicalised individuals are not a homogenous group; thus there is no singular extremist profile or radicalisation pathway. Evidently, the process is increasingly complex and multidimensional.

Sageman's (2008a) 'bunch of guys' theory demonstrates that radicalisation is a by-product of the environment; this can be extended to include virtual environments. This complex radicalisation process does not occur within a vacuum; the pathway to violent extremism is unique to the individual and shaped by the broader social, political and cultural environment. Therefore, the online ecosystem in which they operate can become an influential factor in violent extremism. The following sections will address the nexus of technology and radicalisation to violent extremism utilising Sageman's model. Firstly, the argument shall focus upon the Internet's ability to foster a radical identity. This will be explained through increased exposure to extremist material, anonymity, moral outrage and the process of othering. Secondly, it will demonstrate how the creation of virtual communities and echo chambers consolidates the radicalisation process.

5 Radicalisation and Identity Negotiation

The Internet has provided extremists with a global platform to indoctrinate audiences worldwide. Violent extremist views have always existed; however, when ideology is coupled with technology, the reach achieved is amplified. As Crenshaw (2008) notes, terrorism and extremist views are not new in type but new in scope and lethality. The Internet has enabled radical narratives to be disseminated easily, cheaply, quickly and globally. Therefore due to greater ease of access, more individuals can become exposed to violent extremist content. This in turn increases the risk of ideological indoctrination. Thus, the breadth and depth of online space are inherently problematic.

Anonymity is a significant characteristic of the Internet. Nowadays, technologies for maintaining security and privacy are more readily accessible. For example, this can be seen in the increase in encrypted messaging applications. According to Suler (2004), the perception of anonymity in online communication is an element that may evoke the 'online disinhibition effect'. Extremist Internet users typically view their online identity to be distinct from their physical identity fostering the assumption that virtual actions have lesser repercussions. Thus, the notion that the online sphere guarantees anonymity and less risk increases the likelihood of violent and aggressive rhetoric (Suler, 2004). However, scholars have argued that extremist discourse online does not necessarily correlate with real-life actions. Evidence suggests that violent extremist discourse has become normalised as governments have struggled to distinguish between empty rhetoric and serious intent due to the sheer volume of violent threats (Conway, 2017).

Terrorist scholar Benson (2014) remains sceptical of the nexus between the Internet and radicalisation. Whilst he acknowledges the rapid growth of technology, he maintains that the severity of the threat is frequently overstated in literature. Communication technology has evolved over time; he argues the Internet is merely another development and is not particularly distinct from other technologies. Benson further suggests that in actuality extremist groups such as Al-Qaeda can become hindered by technology. Increased use of the Internet in radicalisation creates a perception of anonymity which can be exploited by intelligence agencies to pinpoint locations of violent extremists. Thus, according to Benson (2014), a large presence on mainstream social media creates avenues for detection and weaknesses within extremist's virtual infrastructure and networks.

In recent times, social media have witnessed an influx of emotionally pervasive propaganda utilising graphic images and videos to spread ideologies. Images are highly accessible in the sense that they are recognisable and not language specific; thus they attain high audience consumption (Baaken & Schlegel, 2017). Online platforms permit extremist groups to control and manipulate their own 'victimhood' narrative. Thus, propaganda is created with the intention to shock and disturb which can have a profound impact upon audiences. According to Sageman (2008b), exposure to extremist materials can spark a sense of moral outrage that shocks viewers into action. This is illustrated in the case of Arid Uka. Arid Uka was an extremist with no affiliation to a broader organisation whose outrage and terrorist action was incited by the consumption of violent videos which he perceived as an attack upon Islam (Baaken & Schlegel, 2017). However, in most cases assessing the scale of exposure to extremist content in conjunction with the impact upon an individual is challenging for researchers. Baaken and Schlegel (2017) claim that individuals search for material and ideologies that resonate with their sense of marginalisation, anger and political grievances. Consequently, extremist material that evokes alienation and anger can incite adherence to violent ideologies. Therefore, the Internet enables individuals to negotiate their personal identity and subsequently create a collective in-group identity rooted in shared grievances and violence.

This collective identity in combination with hostility and alienation within communities breeds an 'us' versus 'them' mentality. As Sageman (2008a) recognises, grievances become interpreted as a wider social concern. An 'us' versus 'them' dichotomy centres around the notion of cultural, religious or ideological superiority over others, also known as the process of othering. Trip et al. (2019) note that this behavioural phenomenon is at the core of extremist ideology. In homogenising individuals into a narrow binary, those who do not belong are ascribed negative, discriminative attributes. The intolerance towards individuals with differing worldviews can lead to the justification of violence against 'them', the existential enemy, to defend their in-group (Trip et al., 2019). Consequently, constructing an 'us' and 'them' distinction imposes stigma upon deviants whilst simultaneously evoking a pressure to adopt and conform to the extremist ideology.

Therefore, immersion and dependence upon extremist ideology for the creation of one's own identity can make it difficult to withdraw from this environment. Trip et al.'s (2019) research is evidenced by several academics, including scholars Nouri and Lorenzo-Dus (2019). Their research revealed that the process of othering was performed by alt-right movements on various social media platforms. Out-groups, typically Muslims, were vilified and dehumanised online utilising subhuman terminology such as 'monkeys' and 'morons'. These groups utilised anti-Muslim rhetoric to legitimise violent extremist ideologies and further polarise communities. As a result, extremist content consumption and networking function to re-shape and construct new-found radical identities. In conclusion, increased exposure to extremist content, anonymity, moral outrage and the process of othering within the online domain all contribute in radicalisation to violent extremism.

6   Radicalisation and Online Social Networks

Within academia there is contentious discussion regarding the extent to which the Internet radicalises individuals. It has become ever more apparent that the Internet plays an important, multifaceted role. It is imperative to note that social media does not solely enable the receival of information; increasingly the Internet has become a place in which individuals can interact, experience community and form strong bonds with others. These communities can reinforce or expand pre-existing extremist ideologies that were first acquired through moral outrage and grievances. Thus, extremists experience a community through the online world which was not feasible decades ago.

Radicalisation, the process by which individuals become extremists, is a social process that has significant real-world effects. The Internet provides a gateway for individuals to create networks and connections and build a sense of group solidarity, an interactive dynamic that would not be possible with the use of one-way pamphlets, magazines and broadcasts. This interactive dimension evokes what Bowman-Grieve (2013) refers to as a 'virtual community'. Virtual communities provide a relatively unregulated space for communication through which values, norms and beliefs can be shared. This provides extremists with a central domain and 'safe haven' to interact freely and express violent views without facing anti-extremist sentiment (Bowman-Grieve, 2013). Extremist virtual networks can provide a sense of belonging and thus are likely to attract those who lack strong social bonds within their offline community. Extremist groups draw individuals into alt-tech platforms and in turn socially isolate individuals within an echo chamber (Bowman-Grieve, 2013). This reinforcement and legitimisation of extremist ideologies within a community can consequently accelerate the radicalisation process. In addition, virtual communities are geographically unbounded. Therefore people can form relationships and develop a community without the need for spatial and temporal convergence.

In contrast to earlier findings, however, Bouhana and Wikstrom (2011) conducted a review of extremist research and concluded that the Internet is not an instrumental factor in radicalisation to violent extremism. They conclude that the online environment rather hinders the development of intimate bonds and communities. Thus, there remains very little agreement within extremist literature. These simplistic narratives can become entrenched within public policy. For instance, the UK Home Affairs Committee (2014, p. 7) claimed that the Internet 'will rarely be a substitute for the social process of radicalisation', therefore suggesting that the Internet is not a social structure. Through the evolution of Web 2.0, the Internet is now an interactive platform that has become a dominant means of interpersonal communication. Thus, governments have come to underestimate social media and the Internet as a tool to transmit violent norms and values.

To understand the way in which extremist individuals use the Internet for social networking and identity negotiation, it is important to consider both echo chambers and filter bubbles. The increased sophistication of mainstream platform mechanisms has raised concerns within academia. It has been suggested that recommendation systems invertedly increase exposure to extremist content; Pariser (2011) termed this phenomenon a 'filter bubble'. A filter bubble is the process whereby automatic algorithms select and present content that is tailored and personalised to individual users based upon past Internet usage and engagement. In this way, online mechanisms filter and limit the material that an individual consumes, thereby curating a 'bubble' of tailored interests and personalised topics. Pariser (2011) argues these pre-selected algorithms prohibit users from exploring new information and ideas as individuals are not presented with diverse information beyond their own political, religious or ideological beliefs (Pariser, 2011). Algorithm selection reinforces the users' existing assumptions and consumption patterns generating a space whereby individuals can interact with those who hold similar belief systems. In addition, research suggests that recommendation algorithms favour extremist content. Reed et al.'s (2019) empirical research regarding radical filter bubbles demonstrates how social media algorithms can encourage extremist networks and discourse. They discovered that when a YouTube user views violent extremist content, the algorithm will further recommend the user extreme content. Therefore, individuals can become confined in a network that is embedded in extremist narratives. Algorithmically driven services play an essential role in the ecology of the Internet and thus potentially perpetuate extremist attitudes and contribute to the radicalisation process.

Filter bubbles and virtual communities provide an avenue for like-minded people to connect and form an echo chamber. An echo chamber is an environment whereby an individual selectively exposes themselves to particular information which amplifies and 'echos' existing beliefs, subsequently disregarding conflicting information. This selective exposure creates the illusion of widespread agreement and potentially distorts perceptions of the underlying reality (Von Behr et al., 2013). Von Behr et al. (2013) conducted interviews with 15 radicalised individuals and found that the Internet had an impact in the radicalisation of all participants. Their findings suggest that the online environment can enhance opportunities to access extremist information and act as an echo chamber. The Internet acts as an echo chamber for violent extremist narratives by providing an opportunity to reinforce radical beliefs and shape identities, thus further radicalising the individual. However, Von Behr et al. (2013) indicate that the Internet merely consolidates the offline process of violent radicalisation as opposed to replacing it. Whilst this study analyses a relatively small sample size and lacks generalisability, it is important to acknowledge the difficulties of participant access in terrorism and extremist studies. The terms filter bubble and echo chamber are often used interchangeably. However, an echo chamber refers to when an individual seeks out information that aligns with already held beliefs; thus it is user-driven, whereas filter bubble refers to when news is automatically filtered out which is exclusively platform-driven. It is currently unclear the extent to which online recommendation systems impact users' choice and vice versa. This distinction is essential as filtering technology itself may play a key role in radicalisation to violent extremism and can provide a vital tool in countering violent extremism.

In summary, increased interactivity and anonymity allow for easy propaganda dissemination which in turn increases exposure to shocking and emotive material. For some, this exposure can lead to moral outrage and the process of othering. Extremist groups exploit this perception of the 'other' to isolate vulnerable individuals from offline peers and place them within an echo chamber to eliminate counter-narratives and disagreement. Therefore, online networks function to sustain and validate extremist identities and thus play an influential role in the process of radicalisation. Radicalisation is facilitated by specific mechanisms of information and communication technology, such as algorithmic filtering and interactive platforms. Consequently, the Internet creates numerous opportunities for radicalisation.

7 Convergence of the Offline and Online Worlds of Violent Extremists

The Internet has increasingly been utilised as a tool to radicalise individuals. Despite this, there is a tendency for literature to either understate or overemphasise the causal role of the Internet in the process of radicalisation to violent extremism (Laqueur, 1999; Burke, 2011; Sageman, 2008b). Research frequently attempts to separate virtual radicalisation and so-called 'real-world' radicalisation. However, it is important to note that radicalisation pathways are diverse and ambiguous; thus they cannot be reduced to an offline/online binary. Activities on the Internet have become interconnected with offline actions to such an extent that there is no longer a clear divide. Whilst this binary analysis persists today, recent research has attempted to look beyond simplistic divisions to explore the relationship between the online and offline domain (Gill et al., 2017; Von Behr et al., 2013; Conway, 2017; Whittaker, 2021).

What occurs online has a tremendous impact upon the outside world. Nonetheless, a clear and causal connection between violent extremist ideologies and online activity is yet to be established in contemporary research. Bouhana and Wikstrom (2011) contend that there is minimal empirical data within radicalisation literature, and those that exist are fundamentally flawed. This is often attributed to the ethical and methodological challenges of researching extremist content. Furthermore, violent extremism and the Internet are both rapidly changing realms; thus they are difficult areas to research simultaneously (Conway, 2017). Bouhana and Wikstrom (2011) suggest that to gain a deeper understanding of the radicalisation process, academics and policy makers must adopt an explanatory approach that considers the interaction between an individual and their social environment.

Whilst this chapter has focussed upon the role of the Internet in violent radicalisation, exploration of the offline dimension is equally important. In order to fully understand online radicalisation, it is crucial to consider both online and offline activities. The Internet may enable radicalisation; however it is rarely the exclusive reason for individuals adopting extremist ideology. Research suggests that whilst the Internet is an influential element in radicalisation, actors' engagement exists in both spheres (Gill et al., 2017). Gill et al. (2017) note that there is an important interaction between offline and online worlds as extremist practices, activities and ideologies are intertwined; thus such attempts to distinguish one domain from the other are of little value. Their study contends that the Internet has afforded greater opportunity for radicalisation and mobilisation; nonetheless online communication does not replace offline interaction (Gill et al., 2017), thereby concluding that radicalisation to violent extremism is cyber-enabled as opposed to cyber-dependant. Whilst Gill et al. (2017) analyse secondary data, it remains one of the few empirically based studies concerning extremist use of the Internet.

The online/offline distinction is exemplified by Valentini et al. (2020) who coined the phrase 'onlife radicalisation'. They maintain that radicalisation occurs from an amalgamation of offline and online spheres. There exists a divide between offline and online worlds but no clear-cut boundary as the domains are intertwined. Additionally, Valentini et al.'s (2020) textual analysis of extremist recruitment videos further demonstrates the presence of 'onlife' radicalisation. However, this analysis presents methodological limitations as the researchers purposefully selected texts that contained a wealth of data, thereby leading to selection bias. Valentini et al. (2020) acknowledge the restrictions of their approach and advocate for further empirical research into the 'onlife' domain. Therefore, whilst findings should be interpreted with caution, this paper provides a significant framework to demonstrate the integration between offline and online domains. This offline/online convergence is further evidenced by Whittaker (2021). Through quantitative and qualitative analysis, he found that in the shift towards Web 2.0, 'going online' is no longer a deliberate choice as the Internet is present in all spheres of life. Whittaker (2021) found that only five terrorist actors originally engaged with extremist material through the Internet. Instead the majority became initially influenced by external offline factors such as friends and family. Therefore whilst the Internet is a significant tool and driver of violent radicalisation, it is uncommon for individuals to be radicalised solely online. Consequently, the Internet in association with offline factors shapes the process of radicalisation to violent extremist.

8 The Birth of the Internet of Things and Terrorism

Following the advent and widespread use of the Internet, there have been rapid advancements in information and communication technology. The Internet of Things (IoT) is an instance of such technologies. The term 'Internet of Things' was coined by Kevin Ashton in 1999 originally to promote RFID technology. However, the idea of connected devices stems from 1832, during which the first electromagnetic telegraph was developed, enabling direct interaction between two machines via the transmission of electrical signals. The real IoT history only began with the creation of the Internet in the late 1960s, when the idea was often referred to as 'embedded internet' or 'pervasive computing'. The term 'Internet of Things' did not become widespread until 2010–2011, and it was only in early 2014 that IoT became mainstream (Lueth, 2014).

The IoT connects anything and everything 'online'. It refers to the interconnection of uniquely identifiable embedded computing devices within the current Internet infrastructure. Whilst some IoT devices are ordinary items with built-in Internet connectivity, others are sensing devices developed exclusively with the IoT in mind. The IoT encompasses technologies such as unmanned aerial vehicles (UAVs), smart swarms, the smart grid, smart buildings and home appliances, autonomous cyber-physical and cyber-biological systems, wearables, embedded digital items, machine-to-machine communications, RFID sensors, context-aware computing, etc. 'Each of these technologies has become a specific domain in its own merit. With the new types of devices constantly emerging, the IoT has almost reached its uttermost evolution' (Montasari & Hill, 2019).

The IoT has brought numerous benefits to society, revolutionising the lifestyles of many individuals living in these societies (Montasari et al., 2020a, b). For instance, the IoT-connected sensors can assist farmers with monitoring their crops and cattle in order to enhance production and efficiency and monitor the health of their herds. Intelligent health-connected devices are used to save or improve patients' lives through wearable devices. However, despite its numerous advantages, the IoT devices simultaneously present many challenges to national security. Although the IoT uses the same monitoring requirements similar to those utilised by cloud computing, it poses more security challenges due to their decentralised nature or issues such as volume, variety and velocity. For instance, a sophisticated terrorist could potentially turn IoT nodes into zombies; intercept and manipulate cardiac devices; perform DDoS attacks; and hack in-vehicle infotainment systems, CCTVs and IP cameras, etc. (Montasari et al., 2020a,b, 2021; Montasari & Hill, 2019). Similarly, unmanned aerial vehicles (UAVs) are particularly appealing to terrorists, insurgents and other militant non-state actors in the same manner that they appeal to state actors. The use of UAVs is an inexpensive method to attack a target without risking personnel (Grossman, 2018). For instance, terrorists could load small, commercially available UAVs with explosives and launch devastating attacks against civilians, military personnel or politicians. This also means that a drone attack avoids provoking a reaction of disgust, shock, anger and other emotions from the public that would often result from a suicide attack. With the rapid expansion of commercially available UAVs, drone bombings will almost probably spread in the same manner as suicide bombings and other forms of terrorism (Grossman, 2018).

We are already witnessing an increasing shift in drone terrorism and insurgent hardware. In 2018, attackers attempted to assassinate the Venezuelan president using a DJI Matrice 600 that can carry a load of more than 5 kg. In Iraq and Syria, ISIS used footage from several drones to coordinate their attacks on army bases. Other images taken by drones have appeared in ISIS recruitment videos intended to appeal to and recruit young men (Grossman, 2018). Western intelligence services have already neutralised numerous drone attacks in the planning phase. For instance, in 2016, in a counterterrorism raid, the UK's security services discovered drone manuals and maps of popular London shopping areas (Grossman, 2018). Similarly, in 2012, the United States charged a man with the declared intention of 'to avenge U.S. drone strikes in Iraq' with attempting to fly explosive-laden drones into buildings in Washington, DC. Likewise, in 2013, demonstrators were able to fly a drone in close proximity of German Chancellor (Grossman, 2018).

These threats will only continue to be further exacerbated with the widespread use of 5G and associated technologies that are increasingly inundating the IoT's and edge ecosystem's security. For instance, by exploiting the speed, scale and processing power that 5G-enabled devices offer, terrorists could potentially carry out catastrophic swarm attacks or to process large volumes of data much faster for their clandestine cyber activities.

9   Conclusion

In conclusion, the evolution of violent extremism, in particular its intersections with the Internet, is evident. However, the relationship between extremism and the Internet is complex and multifaceted and requires a nuanced understanding of both online and offline exposure to violent discourse. In contextualising the history of the Internet and extremist research, examining Sageman's (2008a) four stages of radicalisation and exploring how the online environment accelerate this process; this chapter has illuminated the role of the Internet in radicalisation to violent extremism. Overall, this chapter concludes that violent extremist beliefs are the outcome of offline and online interactions. It is vital to recognise that whilst the Internet is not the root cause of violent extremism, the Internet can be used as a tool of radicalisation to encourage, support and disseminate propaganda.

With the advent of virtual domains, limitations of space and time are no longer a hindrance to the formation of extremist networks. Information and communication technology thus enables individuals and more importantly extremists to amplify bonds, develop virtual communities, establish an echo chamber and amplify grievances. Consequently, the Internet plays a significant role in radicalisation to violent extremism. Furthermore, the increasing misuse of UAVs by terrorists or other non-state actors can have a significant impact on Western national security, hence a cause for concern. Considering the examples of the malicious use of drones provided in this chapter, it has been suggested that Western intelligence services might not be adequately prepared to combat drone terrorism (Grossman, 2018). Therefore, in view of security threats presented by the misuse of UAVs, it might be prudent to develop new policies and regulations that would govern the use of commercial drones by individuals without undermining their civil rights in a democratic society.

References

Baaken, T., & Schlegel, L. (2017). Fishermen or swarm dynamics? Should we understand jihadist online-radicalization as a top-down or bottom-up process? *Journal for Deradicalization, 13*, 178–212.

Benson, D. C. (2014). Why the Internet is not increasing terrorism. *Security Studies, 23*(2), 293–328.

Bouhana, N., & Wikstrom, P. H. (2011). *Al Qa'ida-influenced radicalisation: A rapid evidence assessment guided by situational action theory*. Home Office.

Bowman-Grieve, L. (2013). A psychological perspective on virtual communities supporting terrorist & extremist ideologies as a tool for recruitment. *Security Informatics, 2*(1), 1–5.

Burke, J. (2011). Al-Shabab's tweets won't boost its cause. *The Guardian*. Retrieved December 24, 2011, from https://www.theguardian.com/commentisfree/2011/dec/16/al-shabab-tweets-terrorism-twitter

Burke, J. (2016). The age of selfie Jihad: How evolving media technology is changing terrorism. *CTC Sentinel, 9*(11), 16–22.

Conway, M. (2017). Determining the role of the Internet in violent extremism and terrorism: Six suggestions for progressing research. *Studies in Conflict & Terrorism, 40*(1), 77–98.

Conway, M., Scrivens, R., & Macnair, L. (2019). Right-wing extremists' persistent online presence: History and contemporary trends. *The International Centre for Counter-Terrorism - The Hague, 10*, 1–24.

Crenshaw, M. (2008). The debate over "new" vs. "old" terrorism. In I. A. Karawan, W. McCormack, & S. E. Reynolds (Eds.), *Values and violence. Studies in global justice* (4th ed., pp. 117–137). Springer.

Gill, P., Corner, E., Conway, M., Thornton, A., Bloom, M., & Horgan, J. (2017). Terrorist use of the Internet by the numbers: Quantifying behaviours, patterns & processes. *Criminology and Public Policy, 16*(1), 99–117.

Grossman, N. (2018). Are drones the new terrorist weapon? Someone tried to kill Venezuela's president with one. *The Washington Post*. Retrieved July 19, 2021, from https://www.washingtonpost.com/news/monkey-cage/wp/2018/08/10/are-drones-the-new-terrorist-weapon-someone-just-tried-to-kill-venezuelas-president-with-a-drone/

HM Government. (2015). *Revised Prevent duty guidance: For England and Wales*. Retrieved January 29, 2017, from https://www.safecampuscommunities.ac.uk/uploads/files/2019/03/3799_revised_prevent_duty_guidance_england_wales_v2_interactive.pdf

Home Affairs Committee. (2014). *Counter-terrorism: Seventeenth report of session 2013–14*. The Stationery Office.

Laqueur, W. (1999). *The new terrorism: Fanaticism and the arms of mass destruction*. Oxford University Press.

Lueth, K. L. (2014). Why the Internet of Things is called Internet of Things: Definition, history, disambiguation. *IoT Analytics*. Retrieved July 19, 2021, from https://iot-analytics.com/internet-of-things-definition/

Montasari, R., & Hill, R. (2019). Next-generation digital forensics: Challenges and future paradigms. In *2019 IEEE 12th International Conference on Global Security, Safety and Sustainability (ICGS3)* (pp. 205–212). IEEE.

Montasari, R., Hill, R., Montaseri, F., Jahankhani, H., & Hosseinian-Far, A. (2020a). Internet of things devices: Digital forensic process and data reduction. *International Journal of Electronic Security and Digital Forensics, 12*(4), 424–436.

Montasari, R., Hill, R., Parkinson, S., Peltola, P., Hosseinian-Far, A., & Daneshkhah, A. (2020b). Digital forensics: Challenges and opportunities for future studies. *International Journal of Organizational and Collective Intelligence, 10*(2), 37–53.

Montasari, R., Jahankhani, H., Hill, R., & Parkinson, S. (Eds.). (2021). *Digital forensic investigation of Internet of Things (IoT) devices*. Springer.

Nouri, L., & Lorenzo-Dus, N. (2019). Investigating reclaim Australia and Britain First's use of social media: Developing a new model of imagined political communities online. *Journal for Deradicalization, 18*, 1–37.

Pariser, E. (2011). *The filter bubble: What the Internet is hiding from you*. Penguin Press.

Reed, A., Whittaker, J., Votta, F., & Looney, S. (2019). *Radical filter bubbles: Social media personalization algorithms and extremist content*. Global Research Network on Terrorism and Technology.

Sageman, M. (2008a). *Leaderless Jihad: Terror networks in the twenty-first century*. University of Pennsylvania Press.

Sageman, M. (2008b). The next generation of terror. *Foreign Policy, 16*, 37–42. Retrieved 26 January, 2021, from https://artisinternational.org/articles/Sageman_Next_Generation_of_Terror.pdf

Suler, J. (2004). The online disinhibition effect. *CyberPsychology & Behaviour, 7*(3), 321–326.

Trip, S., Bora, C. H., Marian, M., Halmajan, A., & Drugas, M. I. (2019). Psychological mechanisms involved in radicalization and extremism. A rational emotive behavioral conceptualization. *Frontiers in Psychology, 10*, Article 437.

United Nations. (2016). *Handbook on the management of violent extremist prisoners and the prevention of radicalization to violence in prisons*. United Nations Office on Drugs and Crime.

Valentini, D., Lorusso, A. M., & Stephan, A. (2020). Onlife extremism: Dynamic integration of digital and physical spaces in radicalization. *Frontiers in Psychology, 11*, Article 524.

Von Behr, I., Reding, A., Edwards, C., & Gribbon, L. (2013). *Radicalisation in the digital era: The use of the Internet in 15 cases of terrorism and extremism*. RAND.

Whittaker, J. (2021). The online behaviours of Islamic state terrorists in the United States. *Criminology & Public Policy*, 1–27.

The Impact of the Internet and Social Media Platforms on Radicalisation to Terrorism and Violent Extremism

Kate Gunton

1 Introduction

Violent extremism was documented long before the birth of the Internet in the late 1960s (Navarria, 2016)—as far back as the late nineteenth and early twentieth centuries with the Russian and European anarchists (Aydinli, 2008). The Internet has provided a network of global communication since the late 1980s that can spread information immediately to a global audience (United Nations Office on Drugs and Crime, 2012). Violent extremist groups have exploited online platforms, such as Facebook and Twitter, provided by the Internet to disseminate propaganda, to recruit and radicalise violent extremist actors, to fund their ideologies and to plan potential training and attacks (United Nations Office on Drugs and Crime, 2012). According to Shahar (2007), the Internet has allowed previously isolated and widely distributed extremist groups, such as Jihadists, to create a global network allowing them to become transnational rather than localised to specific countries. The increased online presence of violent extremists indicates that many individuals are more likely to become radicalised by violent extremist ideologies (Neo, 2016). The extent to how much the Internet affects radicalisation has been heavily debated by academics, often referred to as the online and offline dichotomy, and whether radicalisation occurs online or offline (Valentini et al., 2020). Some academics such as Koehler (2014) have argued that the Internet significantly affects the process of radicalisation to violent extremism. Whilst other academics, such as the Soufan Group (2015), argue that radicalisation primarily occurs offline and that the Internet is not a primary factor in violent extremist radicalisation.

K. Gunton (✉)
Hillary Rodham Clinton School of Law, Swansea University, Swansea, UK
e-mail: http://www.swansea.ac.uk

© The Author(s), under exclusive license to Springer Nature Switzerland AG 2022   167
R. Montasari et al. (eds.), *Privacy, Security And Forensics in The Internet of Things (IoT)*, https://doi.org/10.1007/978-3-030-91218-5_8

The remainder of this chapter is structured as follow. Section 2 offers relevant definitions. Section 3 critically discusses the evidence suggesting that the Internet affects radicalisation to violent extremism. Specifically, this section focuses on arguments surrounding echo chambers, opportunities for women to remain anonymous and the role of identity construction for the youth. Section 4 presents a critical discussion on the evidence negating the notion that the Internet can affect radicalisation to violent extremism. This section focuses primarily on offline persuasion and a 'false dichotomy'. Finally, the chapter is concluded in Sect. 5.

2 Definitions

2.1 Violent Extremism

There is no universally agreed definition for violent extremism among policymakers and academics. The UK Government's definition of extremism is the 'vocal or active opposition to fundamental British values, including democracy, the rule of law, individual liberty and mutual respect and tolerance of different faiths and beliefs' (Home Office, 2015). However, this definition has been heavily criticised for being extremely ambiguous as the values are not clearly characterised, which can contribute to the stigmatisation of specific communities (Vincent & Hunter-Henin, 2018). It could also undermine the basic principles of democracy by branding individuals as extremists for having opinions outside of mainstream society (Khan, 2019). Stephens et al. (2019) argue that there are clear distinctions between definitions of extremism that are idealistic, which refers to an ideology that is opposed to a society's values, and behavioural, which refers to an ideology that focuses on the methods actors use to achieve a political objective. Most definitions of 'violent extremism' focus on the behavioural definition in which violent acts are the means to achieve goals compared to more idealistic definitions of 'non-violent' extremism that mainly focus on the extreme belief system itself (Neumann, 2003). It has been argued that distinguishing between non-violent and violent extremism is 'naive and dangerous' as any extremist actor can harbour a belief system without using measures to affect societal change only to turn to violence when deemed necessary (Schmid, 2014, p. 20). Despite this, for the purpose of this chapter, there will be a distinction between non-violent and violent extremism as the discussion is centred specifically around radicalisation to terrorism and violent extremism. As such, the definition of violent extremism that will be used in this chapter is 'encouraging, condoning, justifying, or supporting the commission of a violent act to achieve political, ideological, religious, social, or economic goals' (Federal Bureau of Investigation (FBI), n.d., para. 1).

2.2 Radicalisation

There is a significant amount of conceptual ambiguity and unclarity when defining radicalisation due to a variety of differing definitions, with some highlighting cognitive behaviour and some other extremist thinking (Neumann, 2003). Additionally, there are differences in the conceptualisation of radicalisation in research. Some scholars focus on the process of radicalisation to a radical actor (Schmid, 2013), some place emphasis on the process of radicalisation to an extremist actor (Helfstein, 2012) and others on the process of radicalisation a terrorist actor (Doosje et al., 2016). It is important to recognise that these conceptual ambiguities can substantially impact on approaches taken to counter radicalisation (Neumann, 2003) and can prevent researchers from gaining a holistic understanding of terrorism. For the purpose of this chapter, radicalisation will be defined as the process in which an 'individual adopts an extremist belief system, including the willingness to use, support, or facilitate violence as a method to effect societal change' (Threat of Islamic radicalisation to the homeland, 2007, p. 4). This definition considers both extremist behaviour and extremist thinking which is important when attempting to gain a holistic understanding of violent extremist radicalisation and, ultimately, how it is affected by the Internet. It also conceptualises radicalisation as the process of becoming a violent extremist actor, rather than a radical or terrorist actor, which is central to the discussion within this chapter.

3 Discussion of Supporting Evidence

This section provides a critical discussion of evidence suggesting that the Internet can affect radicalisation to violent extremism.

3.1 Echo Chambers

Some academics argue that the Internet can act as an echo chamber, also referred to as a 'filter bubble' (Reed et al., 2019), that allows individuals to easily access information that will reinforce their beliefs and strengthen them (Briggs, 2014). The Internet has filtering and recommendation software that limits the information that they receive, which enables individuals to self-reinforce their own biases (O'Hara & Stevens, 2015). Prior to the Internet, individuals would have to actively search for information through newspapers or magazines to confirm their views and, in the process, may stumble across information that challenges their biases (O'Hara & Stevens, 2015). Warner (2010) found that individuals' attitudes were easily influenced when exposed to heavily biased news reports and suggested that these individuals' attitudes may harden if they begin to filter out opposing views.

Additionally, Edwards and Gribbon (2013) found that some interviewees expressed that the ease of navigation on the Internet reinforced their existing views when searching for information on religious extremist views. For one interviewee, the ease of access to information on the Internet was the cause of their behaviour change offline as they started to prefer to read extremist material online rather than watching their normal TV shows (Edwards & Gribbon, 2013).

Social media platforms, such as Twitter, enhance the echo chamber effect as violent extremist groups, such as ISIS, can rapidly release propaganda messages to limitless audiences and, ultimately, reinforce their extremist ideologies (Awan, 2017). Due to the nature of social media, individuals may unintentionally encounter extremist propaganda online through trending extremist content or hashtags (Berger, 2016). For example, in October 2015, a small number of white nationalist extremists tweeted in high volumes to amplify their message to boycott *Star Wars: The Force Awakens* as they believed that it promoted anti-white messages, which incited media coverage and increased interest around extremist groups (Berger, 2016). Using Warner (2010)'s argument above, it could be argued that individuals who are constantly exposed to extremist ideologies on social media are highly likely to have their attitudes shaped and hardened by propaganda messaging and could lead to radicalisation. This is supported by Awan (2017)'s study on the role of ISIS on social media as it was found that the presence of ISIS on Twitter and Facebook had acted as an echo chamber whereby ISIS narratives are redistributed and reinforced which could accelerate the online radicalisation process.

However, there is very little evidence to show that the echo chambers on the Internet can affect the radicalisation of an individual (Hussain & Saltman, 2014). Whilst there is a research that demonstrates the existence of echo chambers, it is not often credible (O'Hara & Stevens, 2015). For example, Warner (2010)'s research was based on a small sample of sources which instantly creates issues with representativeness as the evaluated effects cannot be generalised to all forms of news sources. Schlegel (2019) argues that the existence of echo chambers and filter bubbles significantly depends on the design and operation of individual social media platforms and, consequently, cannot be generalised. This is shown by recent research by Reed et al. (2019) that found that users of social media platforms Gab and Reddit were not exposed to more right-wing extremist content after interaction with similar material, which negates the existence of an echo chamber effect. Thus, it can be argued that echo chambers on the Internet provide an environment in which individuals can strengthen and reinforce their beliefs depending on the online platform and, thus, can facilitate rather than directly affect radicalisation to violent extremism.

3.2 Anonymity

The opportunity for anonymity on the Internet has removed barriers for certain groups that exist in the offline world that prevents them from becoming involved in extremism, leading to an increase in online violent extremist radicalisation (Von Behr et al., 2013). For instance, the role of women in Jihadi and Islamic State ideology is based in a home and family environment preventing women from being able to involve themselves in violence (Pearson, 2016). It may be prohibited for some women to seek out information with extremist groups in an offline environment or to express their views in public (Briggs & Strugnell, 2011, as cited in Von Behr et al., 2013). Additionally, the Internet allows women to interact anonymously with violent extremist groups without damaging their position in their own offline community (Sanchez, 2014). Research has shown that women with experiences of rape were more vulnerable to online radicalisation as, in certain communities that hold women to account for their own sexual propriety, the pride related to becoming a member of an extremist group overshadowed the shame of their sexual victimisation (Bloom, 2016).

There have been high-profile cases of female violent extremists who have self-radicalised via the Internet. For example, British student Roshonara Choudhry stabbed her local MP for voting for the Iraq war after viewing violent extremist messaging online (British Broadcasting Corporation (BBC) News, 2010). Pearson (2016) suggests that Choudhry may have adopted an anonymous gender identity through the Internet which allowed her to engage with the violent extremist ideologies of Al-Qaeda, despite their positions of the role of women. Additionally, it was suggested that 'the gradual construction of an online extremist identity appears to have created an . . . dissonance between this and her multiple 'offline' identities' (Pearson, 2016, p. 23). This is supported by the theories of de-individualisation and group polarisation (Spears et al., 1990) which argue that anonymity offered by the Internet leads to a significant amount of identification with the group and, subsequently, results in a lack of self-awareness. Thus, it can be argued that the Internet does affect radicalisation to extremism and, in limited cases, violent extremism. However, it should be noted that more research is necessary to understand the online radicalisation process from women's non-violent extremism to violent extremism (Pearson, 2016).

3.3   Youth and Identity Construction

There has been a substantial amount of research highlighting that the Internet can affect radicalisation to violent extremist groups due to the flexibilities of constructing identities online. For example, Campelo et al. (2018) found that experiences of identity conflict and personal uncertainty, alongside a variety of other individual, societal and environmental factors, make young individuals vulnerable to online radicalisation. Similarly, Lynch (2013) found that young British Muslims are more vulnerable to online radicalisation due to identity conflicts caused by integration and cultural conflict between generations. According to Meeus (2015), religious young individuals with identity conflicts may pursue a potential identity change within a violent extremist ideology, particularly in cases where there is a lack of offline social networks. Research by Cherney et al. (2020) highlighted that the Internet played a significant role in the radicalisation of 21 out of 33 Australian youths—8 of the 21 Australian youths reported to having distant relationships with family members. Thus, it can be argued that violent extremist groups are able to provide individuals with an identity conflict with an opportunity to develop a shared social identity, which can lead to de-individualisation and the adoption of shared extremist ideologies (McKenna & Bargh, 2000).

However, the ability to construct a social identity online is highly dependent on a significant level of trust in social media platforms. Research has found that social media users displayed significant levels of suspicion and mistrust towards certain platforms despite trusting the sincerity of violent extremist propaganda (Hegghammer, 2014, as cited in Beadle, 2017). A survey by the YouGov-Cambridge Globalism Project highlighted that only 83% of British respondents had little, if any, trust in social media platforms, such as Twitter (Hern, 2019). Thus, it can be argued that the Internet is able to affect radicalisation as it gives young individuals opportunities to develop a shared identity, but this depends on the level of trust that they have in the social media platform being used to view the extremist propaganda.

Before moving on to the next part of the discussion, it is essential to acknowledge that most of the research suggesting that individuals may radicalise online is based on little, if any, empirical evidence. Yet, it is difficult to establish a cause and effect when there is no universal agreement into the definitions of extremism and radicalisation due to the lack of consensus on the exact variables that need to be measured. Additionally, as previously mentioned, the empirical data that does exist consists of small and qualitative samples that lack generalisability. For example, in Cherney et al.'s (2020) analysis of Australian youth, the sample consisted of 33 cases of young extremists—of which only one case involved a white supremacist with the rest of the cases being Islamist extremists. Whilst the sample is completely unrepresentative, the researchers do acknowledge that there are limitations in generalising beyond their own study (Cherney et al., 2020).

4 Discussion of Opposing Evidence

This section provides a critical discussion of evidence negating that the Internet can affect radicalisation to terrorism and violent extremism.

4.1 Offline Persuasion

According to the Soufan Group (2015), offline persuasion through the existence of 'hotbeds of recruitment' plays the most significant role in radicalisation. Hotbeds of recruitment refer to locations that have a disproportionate number of Jihadist extremist recruits compared to other locations which, in turn, generates recruitment through a network of personal contacts (Soufan & Schoenfeld, 2016). For example, in the UK, the networks of offline recruiters run by violent extremist Anjem Choudary managed to recruit around 760 members of the Islamic State, with many of them travelling to Syria as a result (Anthony, 2014). Research by Schils and Verhage (2017) found that many respondents reported that their extremist ideology was mostly influenced by their social environments, such as family and peers. The involvement of a known contact such as a family member or friend in the process of radicalisation is more likely to encourage an individual to join the Islamic State compared to radicalisation through social media (Soufan Group, 2015). Groups of young individuals vulnerable to radicalisation within these hotbeds of recruitment are provided with a sense of belonging and purpose which, then, leads to the radicalisation of groups of friends, families and neighbours as they convince each other to join the Islamic State together (Soufan Group, 2015).

Much of the evidence demonstrating the hotbeds of recruitment comes from inside prisons. Due to the number of foreign fighters that have been imprisoned before joining a violent extremist group (Renard et al., 2018), it seems inevitable that some of them were radicalised, or had their extremist views strengthened, whilst in prison. For example, in 2014, former ISIS soldier Mehdi Nemmouche murdered four people in Brussels just 2 years after he was released from a 5-year prison sentence, where he became radicalised by associating himself with violent Islamist inmates (Counter Extremism Project, n.d.-a). Similarly, Amedy Coulibaly, who was responsible for killing four people in the 2015 Paris' supermarket attack, was exposed to violent extremist ideology through associating with an al-Qaeda recruiter (Counter Extremism Project, n.d.-b). Regarding evidence supporting offline persuasion compared to persuasion on the Internet, Schils and Verhage (2017) discovered that a large number of extremist supporters on social media platforms are not easily persuaded into offline networking. Thus, it could be indicated that, despite the Internet's ability to reach a limitless audience, offline persuasion through hotbeds of recruitment has a greater effect on the process of radicalisation for those with opportunities to access extremism in the physical world. Although there are some issues with the lack of empirical data and generalisability in research on offline persuasion and hotbeds of radicalisation, it does not appear to affect the validity of the argument based on the real-world examples of radicalisation in prisons.

4.2 False Dichotomy

Gill et al. (2015) argue that there is a 'false dichotomy' that unrealistically separates the process of online and offline radicalisation as most individuals radicalised are affected by both equally. Each case of radicalisation is completely unique depending on factors such as the age and gender of the radicalised extremist, which means there are no generalisable variables to measure or standardised process to radicalisation (Hoskins & O'Loughlin, 2009). Rather than treating both online and offline radicalisation as separate processes, Valentini et al. (2020) argue that they should be incorporated into one model as online and offline experiences should both be taken into account. Research on lone actors by Lindekilde et al. (2019) has shown that the processes of online and offline radicalisation often occur simultaneously and are interconnected. For example, Gill et al. (2017) found that those radicalised online were more than 4% more likely to have networked in an offline environment compared to those who were not radicalised online and were more than 3% more likely to experience an offline interaction with violent extremist. Koehler (2014) came to similar conclusions arguing that online and offline patterns of radicalisation are interconnected, particularly in terms of strengthening ideologies and propaganda dissemination. For Scrivens et al. (2020), it is essential that future research on violent extremist radicalisation focuses on combining online and offline data to understand the interconnectivity between the two.

However, despite the suggestions for research to incorporate online and offline radicalisation into an integrated model, Saifudeen (2014) argues that it would be more helpful to fully explore the role of the Internet in radicalisation rather than just reducing its role to a facilitator. Additionally, Szmania and Fincher (2017) argue that the difficulty in establishing an online vs. offline dichotomy highlights a lack of understanding of radicalisation rather than an occurrence of a 'false dichotomy'. Thus, it could be argued that viewing online and offline radicalisation as an integrated model may be able to provide a holistic understanding of extremist radicalisation. Yet, it would not be effective as there is a lack of understanding and empirical research around online and offline radicalisation as separate entities.

5 Conclusion

Having critically discussed how the Internet affects radicalisation to violent extremism, it can be concluded that the Internet mostly increases opportunities for women and young individuals to become radicalised in ways that may not be possible offline. It was demonstrated that the opportunity for anonymity allows women to access information that would not be available to them in offline contexts due to the role of women in violent extremist groups, such as Al-Qaeda. Then, it was highlighted that young individuals with identity conflicts and a lack of social network are provided with opportunities to construct a shared extremist identity through the Internet. Although these two arguments presented methodological issues and gaps in research, it was clear that there were direct links between the Internet and radicalisation to violent extremist groups. However, it should be noted that the online radicalisation process from women's non-violent extremism to violent extremism needs more research in the future (Pearson, 2016). In the discussion supporting how the Internet affects radicalisation, it was also argued that evidence demonstrated the existence of echo chambers on the Internet, but that it does not directly affect radicalisation to violent extremism; rather it strengthens extremist beliefs if frequently exposed to extremist messaging.

In the critical discussion of evidence negating that the Internet can affect radicalisation to violent extremism, it was argued that offline persuasion through hotbeds of recruitment significantly affected the likelihood of radicalisation to violent extremism for individuals who had an opportunity to access extremism in the physical world, such as prisons. The issues of generalisability and lack of empirical data for the hotbed of recruitment argument were acknowledged, but the validity of the argument was strengthened by the real cases of violent extremist radicalisation in prisons. It is also important to note that offline persuasion would have less of an effect on the radicalisation of women as they would not be able to engage in some of the same opportunities as men in the physical world. When critically discussing the evidence suggesting the existence of a 'false dichotomy', it was argued that researching online and offline radicalisation as an integrated model would not be effective when there is already a lack of empirical research around online radicalisation. Throughout the chapter it was acknowledged that most of the research suggesting that individuals may be radicalised online is based on little, if any, empirical evidence. To conduct future objective research to establish a cause and effect, it would be necessary to develop universally agreed definitions for extremism and radicalisation. Additionally, future research may need to fill in gaps of understanding around whether echo chambers could directly affect radicalisation and the understanding around whether offline persuasion affects radicalisation outside of hotbeds of recruitment.

References

Allen, C. E. (2007). Threat of Islamic radicalization to the homeland. In *Testimony before the US Senate Committee on Homeland Security and Government Affairs* (p. 4).

Anthony, A. (2014). Anjem Choudary: The British extremist who backs the caliphate. *The Guardian.*. Retrieved from https://www.theguardian.com/world/2014/sep/07/anjem-choudary-islamic-state-isis

Awan, I. (2017). Cyber-extremism: Isis and the power of social media. *Society, 54*(2), 138–149.

Aydinli, E. (2008). Before jihadists there were anarchists: A failed case of transnational violence. *Studies in Conflict & Terrorism, 31*(10), 903–923.

Beadle, S. (2017). *How does the Internet facilitate radicalization?* (pp. 1–19). War Studies Department, King's College.

Berger, J. M. (2016). Nazis vs. ISIS on Twitter: A comparative study of white nationalist and ISIS online social media networks.

Bloom, M. (2016). The changing nature of women in extremism and political violence. *Freedom from Fear, 11*, 40–54.

Briggs, R. (2014). *Policy briefing: Radicalisation, the role of the Internet.* Institute for Strategic Dialogue. Retrieved from https://www.isdglobal.org/wp-content/uploads/2016/07/StockholmPPN2011_BackgroundPaper_FOR20WEBSITE.pdf

Briggs, R., & Strugnell, A. (2011). Radicalisation: The role of the internet. Policy Planners Network Working Paper. Institute for Strategic Dialogue.

British Broadcasting Corporation News. (2010). *Student guilty of attempted murder of MP Stephen Timms.* Retrieved from https://www.bbc.co.uk/news/uk-england-london-11673616

Campelo, N., Oppetit, A., Neau, F., Cohen, D., & Bronsard, G. (2018). Who are the European youths willing to engage in radicalisation? A multidisciplinary review of their psychological and social profiles. *European Psychiatry, 52*, 1–14.

Cherney, A., Belton, E., Norham, S. A. B., & Milts, J. (2020). Understanding youth radicalisation: An analysis of Australian data. *Behavioral Sciences of Terrorism and Political Aggression*, 1–23.

Counter Extremism Project. (n.d.-a). *Mehdi Nemmouche*. Retrieved from https://www.counterextremism.com/extremists/mehdi-nemmouche

Counter Extremism Project. (n.d.-b). *Amedy Coulibaly*. Retrieved from https://www.counterextremism.com/extremists/amedy-coulibaly

Doosje, B., Moghaddam, F. M., Kruglanski, A. W., De Wolf, A., Mann, L., & Feddes, A. R. (2016). Terrorism, radicalization and de-radicalization. *Current Opinion in Psychology, 11*, 79–84.

Edwards, C., & Gribbon, L. (2013). Pathways to violent extremism in the digital era. *The RUSI Journal, 158*(5), 40–47.

Federal Bureau of Investigation. (n.d.). *What is violent extremism?* Retrieved from https://www.fbi.gov/cve508/teen-website/what-is-violent-extremism

Gill, P., Corner, E., Conway, M., Thornton, A., Bloom, M., & Horgan, J. (2017). Terrorist use of the Internet by the numbers: Quantifying behaviors, patterns, and processes. *Criminology & Public Policy, 16*(1), 99–117.

Gill, P., Corner, E., Thornton, A., & Conway, M. (2015). *What are the roles of the Internet in terrorism? Measuring online behaviours of convicted UK terrorists.* Australian National University.

Hegghammer, T. (2014). Interpersonal trust on Jihadi internet forums. Norwegian Defence Research Establishment, 1–43.

Helfstein, S. (2012). *Edges of radicalization: Ideas, individuals and networks in violent extremism.* Military Academy West Point NY Combating Terrorism Center.

Hern, A. (2019). Britons less trusting of social media than other major nations. *The Guardian*. Internet.

Home Office. (2015). *Revised prevent duty guidance for England and Wales.* Home Office.

Hoskins, A., & O'Loughlin, B. (2009). Media and the myth of radicalization. *Media, War & Conflict, 2*(2), 107–110.

Hussain, G., & Saltman, E. M. (2014). Jihad trending: A comprehensive analysis of online extremism and how to counter it. *Quilliam*.

Khan, S. (2019). *Challenging hateful extremism.* Commission for Countering Extremism.

Koehler, D. (2014). The radical online: Individual radicalization processes and the role of the Internet. *Journal for Deradicalization, 1*, 116–134.

Lindekilde, L., Malthaner, S., & O'Connor, F. (2019). Peripheral and embedded: Relational patterns of lone-actor terrorist radicalization. *Dynamics of Asymmetric Conflict, 12*(1), 20–41.

Lynch, O. (2013). British Muslim youth: Radicalisation, terrorism and the construction of the "other". *Critical Studies on Terrorism, 6*(2), 241–261.

McKenna, K. Y., & Bargh, J. A. (2000). Plan 9 from cyberspace: The implications of the Internet for personality and social psychology. *Personality and Social Psychology Review, 4*(1), 57–75.

Meeus, W. (2015). Why do young people become Jihadists? A theoretical account on radical identity development. *European Journal of Developmental Psychology, 12*(3), 275–281.

Navarria, G. (2016). How the Internet was born: From the ARPANET to the Internet. *The Conversation*. Retrieved from https://theconversation.com/how-the-Internet-was-born-from-the-arpanet-to-the-Internet-68072

Neo, L. S. (2016). An Internet-mediated pathway for online radicalisation: RECRO. In M. Khader, L. S. Neo, G. Ong, E. T. Mingyi, & J. Chin (Eds.), *Combating violent extremism and radicalisation in the digital era* (pp. 197–224). IGI Global.

Neumann, P. R. (2003). The trouble with radicalization. *International Affairs, 89*(4), 873–893.

O'Hara, K., & Stevens, D. (2015). Echo chambers and online radicalism: Assessing the Internet's complicity in violent extremism. *Policy & Internet, 7*(4), 401–422.

Pearson, E. (2016). The case of Roshonara Choudhry: Implications for theory on online radicalization, ISIS women, and the gendered jihad. *Policy & Internet, 8*(1), 5–33.

Reed, A., Whittaker, J., Votta, F., & Looney, S. (2019). *Radical filter bubbles: Social media personalization algorithms and extremist content.* Global Research Network on Terrorism and Technology.

Renard, T., Coolsaet, R., Heinke, D. H., Malet, D., Minks, S., Raudszus, J., & Van Ginkel, B. (2018). *Returnees: Who are they, why are they (not) coming back and how should we deal with them?: Assessing policies on returning foreign terrorist fighters in Belgium, Germany and the Netherlands* (Vol. 101). Egmont-Royal Institute for International Relations.

Saifudeen, O. A. (2014). *The cyber extremism orbital pathways model.* S. Rajaratnam School of International Studies, Nanyang Technological University.

Sanchez, S. E. (2014). *The internet and the radicalization of Muslim women.* Western Political Science Association.

Schlegel, L. (2019, September 19). Chambers of secrets? Cognitive echo chambers and the role of social media in facilitating them. *European Eye on Radicalisation.* https://eeradicalization.com/echo-chambers-social-media-schlegel/

Schils, N., & Verhage, A. (2017). Understanding how and why young people enter radical or violent extremist groups. *International Journal of Conflict and Violence, 11*(2).

Schmid, A. P. (2013). Radicalisation, de-radicalisation, counter-radicalisation: A conceptual discussion and literature review. *ICCT Research Paper, 97*(1), 22.

Schmid, A. P. (2014). Violent and non-violent extremism: Two sides of the same coin. *ICCT Research Paper*, 1–29.

Scrivens, R., Gill, P., & Conway, M. (2020). The role of the internet in facilitating violent extremism and terrorism: Suggestions for progressing research. In *The Palgrave handbook of international cybercrime and cyberdeviance* (pp. 1417–1435). Springer.

Shahar, Y. (2007). The Internet as a tool for intelligence and counter-terrorism. In B. Ganor, K. Von Knop, & C. Duarte (Eds.), *Hypermedia seduction for terrorist recruiting* (pp. 140–153). IOS Press.

Soufan, A., & Schoenfeld, D. (2016). Regional hotbeds as drivers of radicalization. In *Jihadist hotbeds: Understanding local radicalisation processes.* ISPI.

Soufan Group. (2015). *Foreign fighters: An updated assessment of the flow of foreign fighters into Syria and Iraq.* Soufan Group.

Spears, R., Lea, M., & Lee, S. (1990). De-individuation and group polarization in computer-mediated communication. *British Journal of Social Psychology, 29*(2), 121–134.

Stephens, W., Sieckelinck, S., & Boutellier, H. (2019). Preventing violent extremism: A review of the literature. *Studies in Conflict & Terrorism*, 1–16.

Szmania, S., & Fincher, P. (2017). Countering violent extremism online and offline. *Criminology & Public Policy, 16*(1), 119–125. https://onlinelibrary.wiley.com/doi/epdf/10.1111/1745-9133.12267

United Nations Office on Drugs and Crime. (2012). *The use of the Internet for terrorist purposes.* Retrieved from https://www.unodc.org/documents/frontpage/Use_of_Internet_for_Terrorist_Purposes.pdf

Valentini, D., Lorusso, A. M., & Stephan, A. (2020). Onlife extremism: Dynamic integration of digital and physical spaces in radicalization. *Frontiers in Psychology, 11*, 524.

Vincent, C., & Hunter-Henin, M. (2018). The trouble with teaching 'British values' in school. *Independent.* Retrieved from https://www.independent.co.uk/news/education/british-values-education-what-schools-teach-extremism-culture-how-to-teachers-lessons-a8200351.html

Von Behr, I., Reding, A., Edwards, C., & Gribbon, L. (2013). *Radicalisation in the digital era: The use of the Internet in 15 cases of terrorism and extremism.* Retrieved from https://www.rand.org/content/dam/rand/pubs/research_reports/RR400/RR453/RAND_RR453.pdf

Warner, B. R. (2010). Segmenting the electorate: The effects of exposure to political extremism online. *Communication Studies, 61*(4), 430–444.

The Internet, Social Media and the Internet of Things in Radicalisation to Terrorism and Violent Extremism

Megan Thomas-Evans

1 Introduction

The Internet plays a central role in everyday life, which allows a vast majority of the population to be connected instantaneously, regardless of geographical location (Stewart & Thompson, 2002). The Internet has risen exponentially since the 1990s and today provides terrorists and extremists the capabilities of committing serious threat and attacks on the surrounding population which was not available to them pre-1990 (Salahuddin & Alam, 2015). Weimann (2004) states there are six ways how the Internet facilitates violent extremism.

One way in which the Internet is utilised in violent extremism is radicalisation. To discuss the concept of radicalisation more effectively, there must be a clear definition. However, currently, a universal definition does not exist, making it difficult to enforce international agreements when tackling radicalisation and extremism (Ganor, 2002). Despite the absence of a universal, agreed-upon definition, radicalisation can be described as a process in which a person adopts extremist views and progresses towards engaging in violent behaviours (Hardy, 2018). This is in line with The Prevent Strategy policy's (Home Office, 2011, p. 108) definition of radicalisation. According to The Prevent Strategy (Home Office, 2011), radicalisation is the 'process by which a person comes to support terrorism and forms to extremism leading to terrorism'. The remaining ways in which the Internet can be used to facilitate extremism (Weimann, 2004) will be discussed later in this chapter.

While there are various forms of extremism such as domestic (campaigning for animal rights) and non-violent extremism, this chapter will focus only on

M. Thomas-Evans (✉)
Hillary Rodham Clinton School of Law, Swansea University, Swansea, UK
e-mail: http://www.swansea.ac.uk

violent extremism. The British government define extremism as a 'vocal or active opposition to fundamental British values, including democracy, the rule of law, tolerance of different faiths and beliefs, individual liberty and mutual respect' (Home Office, 2011, p. 107). Extremists can be characterised as political actors who ignore the rule of the law and reject diversity within societies (Schmid, 2013). Striegher (2015, p. 79) states that the Crown Prosecution Service (CPS) defines violent extremism as the 'demonstration of unacceptable behaviour by using any means or medium to express views which foment, justify or glorify terrorist violence in furtherance of particular beliefs'. This is including those who engage in terrorist or criminal violence based on ideological, political or religious beliefs and foster hatred that leads to violence. Violent extremist behaviours often lead to the act of terrorism (Trip et al., 2019), another term which fails to be universally defined, but cannot be ignored when discussing violent extremism.

Most experts and researchers agree on the definition given in the Terrorism Act 2000 (TACT 2000) as an 'action that . . . causes serious violence to a person/people; causes serious damage to property; or seriously interferes or disrupts an electronic system'. The terrorist's threat is to influence the government or inject fear into the public society, with an aim to change a political, religious or ideological belief (Home Office, 2011, p. 108). This chapter will be examining the role of the Internet in radicalisation to violent extremism and will discuss the radicalisation processes individuals experience. It will also address how it is possible for an individual to become radicalised into violent extremism entirely through the Internet without offline communication and physical contact in the outside world. Furthermore, AI and ML technologies will be discussed, specifically how they can be used to decrease extremist material online, which in turn should decrease the number of individuals who become radicalised through social media.

The remainder of this chapter is structured as follows. Section 2 describes terrorism as a process. Section 3 discusses phases of radicalisation, while Sect. 4 examines phases of online terrorism. Sections 5 and 6 critically investigate the roles that SM and the IoT play in enabling terrorism, respectively. Section 7 discusses some of the benefits of the IoT technology in moderating cyberterrorism. Finally, the chapter is concluded in Sect. 8.

2 Terrorism as a Process

Along with the definition previously mentioned, radicalisation can also be explained as a gradual process, whereby individuals are radicalised by friends offline through the Internet, or if they have direct communication with extremist groups (Christmann, 2012). How many steps are in this process, however, differs between various researchers (Gill, 2007; Sageman, 2004; Silber et al., 2007). Sageman's 'bunch of guys' theory (2004) is a group-based social psychological process theory of extremism which includes four steps that a person must encounter to become radicalised: a sense of moral outrage, developing a specific worldview, resonating that

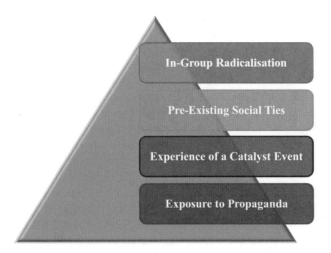

Fig. 1 Steps of radicalisation to terrorism and violent extremism as described by Gill (2007)

worldview with personal experiences and mobilising through interactive networks. The Internet could be involved in all steps towards radicalisation. A moral outrage and the specific worldview could be derived from information seen online in current news articles which encourages the reader to form an opinion based on what they have seen. It could be related to a past personal experience that is sensitive to the individual which influences them to engage further to gain understanding. The Internet would play a role in the final step of Sageman's radicalisation process as individuals mobilise through interactive networks online within chatrooms and social media platforms.

The Internet is very accessible, where individuals can communicate with those who obtain similar views to them which could contribute to their radicalisation process (D'Souza, 2015; Jones, 2009). This is not as easy in the offline world as social groups are limited to external environments such as location of the individual's home. However, social groups created online can lead to social events in the offline world. Groups sharing the same ideology online can meet in a large number which could potentially lead to protest marches in cities, possibly causing disturbances. Gill (2007) also describes radicalisation as a pathway to terrorism and extremist behaviours in four steps as represented in Fig. 1. These include exposure to propaganda, the experience of a catalyst event, pre-existing social ties which aid recruitment and in-group radicalisation.

Propaganda is defined as a coordinated attempt to influence others to actively spread a point of view with the aim to change society's views (Smith, 2020). Propaganda can be viewed online through social media and in many forms of multimedia such as online magazines, videos and images (Weimann, 2014). Social ties could come from current social groups the individual associates themselves with on social media (Jones, 2018). In-group radicalisation could be achieved through online communities where individuals share common beliefs; again, this is less likely to occur in the offline world due to external environments (Torok, 2013), but is not limited to meeting with social media friends in the 'real' world. Gill allows exposure from external sources to be present, whereas Sageman (2004) implies a sense of moral outrage has come from a personal opinion. Although Sageman's (2004) sense of moral outrage may be derived from viewing a news article or video, Gill (2007) specifies radicalisation to occur via exposure to propaganda. Comparing the two theories further, Sageman (2004) presents the theory in a more internal method compared to Gill (2007). Gill's (2007) mention of experience of a catalyst event, past social ties and in-group radicalisation all deter to external environmental factors compared to Sageman's (2004) worldview, resonating it with personal experiences and mobilising through networks which imply more internal views. This could be reflected in the individuals who are radicalised via Sageman's (2004) theory moving to online radicalisation and completing further research via SM and those who are radicalised via Gill's (2007) theory being more extroverted and seeking socialisation from the outside world. This could suggest that those radicalised through the theory of Sageman (2004) are more likely to be a lone actor in terrorist activities and Gill's (2007) more likely to act within a group. A theological psychological approach to the radicalisation process shared by Sageman (2008) and Wiktorowicz (2004) explained how radical religious beliefs are stimulated by group dynamics and lead individuals to become extremists (Kundnani, 2012). These beliefs are derived from the echo chamber theory which explains individuals agree with the norms of a group. Individuals decide to listen to opinions they agree with which create the group dynamics (Sageman, 2008). The Internet acts as an echo chamber on SMPs as following a certain group instead of multiple groups can feed certain information (Ramakrishna, 2011). This perspective influenced Silber et al. (2007), senior intelligence analysts at New York Police Department, to compile the 'NYPD model', describing four stages of radicalisation: pre-radicalisation, self-identification, indoctrination and jihadisation. The Internet could be used and may be a common factor in all steps of this NYPD model into radicalisation through propaganda, social media, videos online and chatrooms.

3   Phases of Radicalisation

Based on the existing studies (Doosje et al., 2016), radicalised individuals follow three phases during the radicalisation process, including sensitivity phase, group membership phase and action phase. The sensitivity phase is being the time when they engage in radical beliefs and the group membership phase where individuals communicate with those who share common beliefs. The action phase is where the radicalised individual participates in radical behaviours and act on their beliefs. Alike the other suggested radicalisation processes, these steps can occur online in the same ways (van den Bos, 2018). Whether these phases will be reached by the individual is dependent on three levels within the sensitivity phase: micro (individual), meso (group) and macro (societal) level (Doosje et al., 2016). The micro level is where the individual seeks for information when they feel a loss of identity and belonging (Baumeister & Leary, 1995). This is likely to occur in the online world as these feelings of significance can be restored by groups such as ISIS by giving new recruits respect and a sense of belonging within the group (Doosje et al., 2016). It could be suggested that those radicalised through Sageman's (2004) theory are more likely to experience this stage due to seeking for significance on SM. This is known as the social identity theory where they have been unable to seek for their identity elsewhere on the outside world (Tajfel & Turner, 1986). At this stage, extremist groups such as ISIS motivate recruits to strengthen their identity with the group, enabling them to adopt norms and values (Hogg et al., 2013). As these extremist groups present a strong structure (Hussin, 2018), someone who experiences personal uncertainty would be attracted to their ideology (Hogg, 2020).

The meso level consists of the external environment the radicalised individual would be surrounded by such as friends, family and those active in extremist groups (Doosje et al., 2016). Research shows it is more like for individuals to become radicalised if peers such as friends and family are participating in learning ideologies too (Radicalisation Awareness Network, 2017). Gill's (2007) theory of radicalisation suggests this, as pre-existing social ties could become a factor of radicalisation. A factor within this level is known as fraternal relative deprivation, where people believe their group has been treated negatively compared to other groups (Crosby, 1976). In context, radical right groups such as the British National Party (BNP), the English Defence League (EDL) and Britain First believe national citizens are treated worse than immigrants (Doosje et al., 2012) and therefore want to highlight and make a change within society where they would perceive they are treated equally or even above them. In the past on SM, the BNP made themselves very public with their views, such as their belief that immigrants were taking 'white British' jobs (Goodwin, 2011). These views could have been spread in the offline world; however, they would not have gained as much publicity. Unless the party travelled throughout the UK to promote their views, they may not have spread further than a county or city. This is how SM plays such a large impact on group policies and beliefs being spread.

The macro level sees the radicalisation process being influenced by societal factors (Moghaddam et al., 2016). The macro level can be described by the effect of globalisation and modernisation (Crenshaw, 1981) along with issues of foreign policy. Due to these factors, it is deemed that globalisation initiates terrorism and violent extremism (Moghaddam, 2008), and even more so using the Internet. For example, with technology live streaming, the 9/11 attacks conducted by al-Qaeda, the terrorist organisation received worldwide recognition from the media (Grusin, 2010). Along with the Christchurch shootings being livestreamed on Facebook, this strongly shows how SM needs to become stronger in regulating terrorist material. Micro, meso and macro factors have a major influence on whether individuals will continue to become radicalised and all interconnect with each other. Once these stages have been experienced by the radicalised individual, the group membership and action phase can take place (Doosje et al., 2016).

Alternative views suggest an individual can foster radicalised beliefs either by having adopted few radical ideas themselves or not having extremist views at all at the beginning of the radicalisation process (Neo et al., 2017). Information they view online can influence them to undertake extremist views, and the Internet is a tool that provides greater opportunity for planning attacks and violent radicalisation (Gill et al., 2014). It must be noted that not everyone who agrees with radical views engages in the physical acts of committing a terror attack (Bertram, 2015). When not actively participating in these events, the radicalised individuals would take on another key role in the organisation of a terror attack such as motivators at events and propaganda distributors which encourage more individuals to join or run social media to promote the organisation (Baugut & Neumann, 2020). Although there is no psychological profile that matches all extremists, research shows those who face socioeconomic disadvantage, government oppression and mental health are more likely to engage in extremist behaviours (Butler et al., 2003; Hudson, 1999).

Radical groups share common beliefs (Borum, 2014) such as believing in serious issues within society and have an urge to change these. The issues can be explained through an overlap of perspectives that draw upon psychological, social, political and economic factors. Political issues such as government oppression can be the root of an extremist ideology (Wibisono et al., 2019). Anger can stem from institutions not dealing with grievances, resulting in engaging in violent behaviours (McCauley & Moskalenko, 2008). As aforementioned, a radical group may arrange march protests via online groups which gains members and influences their beliefs onto others. This is known as deindividuation where individuals behave differently in groups and do not view themselves as individuals; they adopt the opinions of the groups as opposed to developing ideas themselves (Diener, 1979).

This also agrees with Gill's (2007) theory as opinions are shared with an attempt to radicalise individuals further. Individuals would engage in impulsive, deviant and violent acts where they believe they cannot be personally identified (Douglas, 2010), in this instance, participating in violent extremist behaviours such as protests. Details about marches can be posted online and spread to thousands of users for free and at a quick rate (Baruah, 2012). Without the Internet, distribution of details would occur through word of mouth and physical propaganda such as posters, newspapers and TV coverage (Jowett & O'Donnell, 2018). The Internet makes the radicalisation process much more cost-effective and efficient to radical groups by enabling them with a global reach of audience (United Nations Office on Drugs and Crime, 2012). It must be stated that researchers found more Internet research into extremist behaviour, resulting in more physical contact with fellow extremists such as attending marches (Gill et al., 2014).

4   Phases of Online Terrorism

There are three phases of online terrorism which contributes to violent extremism—the early years (1990, 2006), 'Web 2.0' and the regulatory fight era. Tim Berners-Lee invented the World Wide Web in 1989 which led to 16 million users, 0.4% of the world's population, by 1995 (Statista, 2020). Recent figures by Statista

(2020) show active online users have increased to 4.66 billion, covering 59% of the world's population. Within the first phase, terrorists were early adopters of the Internet (Tzezana, 2016). They were attracted to cyber space as it is inexpensive and developed an increased reach of audience (Cohen-Almagor, 2005). The 1990s saw the rise of the radical right, extremist groups with policies leaning towards conservatism, nationalism and anti-immigration. This timeframe also matched the rise of the Internet for the radical right group, the British Nationalist Party (BNP). They used the Internet to their advantage to gain their following on Facebook. The second phase (2007–2015) saw the introduction of social media and mobile technology, which coincides with the rise in ISIS (Patrikarakos, 2018).

Platforms such as YouTube and Twitter saw the first of extremist activity, with around 22,000 Twitter users contributing to support or propaganda distribution for ISIS (Benigni et al., 2017). Onwards from 2016 saw the beginning of phase 3 where platforms took an active approach in removing extremist content and groups on social media, which led to ISIS degrading. Twitter suspended more than 200,000 extremist accounts in 2016, leading to the rise of extremist content appearing on end-to-end encrypted messaging platforms such as Telegram (D'Incau & Soesanto, 2017). This era also saw the devolvement of the BNP, EDL and Britain First, as they became permanently banned from Facebook in 2019 due to falling under their 'dangerous individuals and organisations' policy (Facebook Community Standards, n.d.; Vincent, 2019). The Internet has evolved extremely quickly over the last decade, providing extremist groups with the facilities and materials they require to commit to their success.

According to Weimann (2004), there are six ways through which the Internet can be exploited to facilitate violent extremism. These include recruitment, socialisation, communication, networking, mobilisation and coordination, as represented in Fig. 2.

Recruitment is successful to extremist organisations by using social media through the Internet (Amedie, 2015). Those seeking for new members to join their group exploit existing grievances in vulnerable users (Speckhard & Ellenberg, 2020). They reach out to individuals suggesting they can provide a sense of belonging and a positive life within extremist groups such as the ISIS community. This would be attractive for users who believe they lack a sense of belonging in their community (Dekel & Nuttman-Shwartz, 2009). It can be said that many people have been radicalised and encouraged by propagandists who are overseas and, therefore, would have had to occur online (Home Office, 2011). It is evident that this would not have occurred if the radicalised individual did not view propaganda online or did not have Internet access at all. However, the Home Office (2011, p. 13) states evidence suggests some individuals who have been radicalised in the UK had participated in extremist organisations in the past. Therefore, it may be easier for propagandists to entice those with a history of extremist behaviours to join their organisation. Weimann (2004) expresses socialisation, the process of internalising the norms of a group, is also used through the Internet by violent extremists. Through the process of socialisation, Weimann (2004) reveals users learn the language of the culture, their roles and responsibilities in life and what is expected from them. It

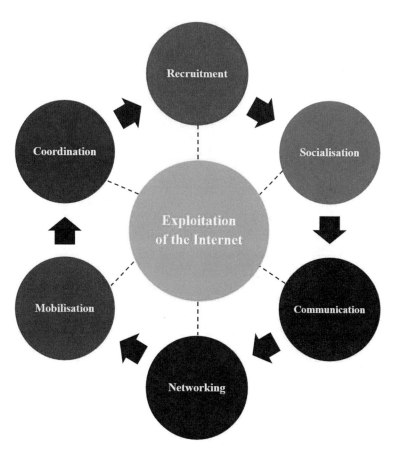

Fig. 2 Six phases through which the Internet can be exploited, as suggested by Weimann (2004)

is possible for group polarisation to occur during the socialisation process. Group polarisation is referred to a group making decisions less rationally and therefore more extreme than initial ideas of individuals within the group (Spears et al., 1990). Users may develop a strong dedication to the online community and withdraw from offline peers if they are socially isolated as they now feel they are a part of a new culture. Socialisation can occur solely online, as the Internet is a safe space for those normally isolated from one another and online interactions fulfil a sense of community (Koehler, 2014). In external environments, it is difficult to find a large group of people who hold the same beliefs in this context. As the Internet provides individuals with instant messaging and reaches to all over the world, it gives them the opportunity to find belonging in groups alike.

Weimann also discusses communication as a vital component of terrorist activity online (2004). Communication is mainly heightened within extremist groups in social media to circulate propaganda, recruiting members, raising funds, attempting to normalise extremist views, advising members on how to support the group and gaining publicity (Archetti, 2013). Organised groups use smaller alt-tech platforms to coordinate mainstream campaigns (Weimann, 2016). Examples of these alt-tech platforms are Gab and Telegram that impose less strict content moderation rules. Gab is a social media platform known for its far-right userbase and free speech capability (Donovan et al., 2018). Telegram, on the other hand, is an instant messaging service where messages are heavily encrypted and can be deleted as soon as it meets the recipient (Anglano et al., 2017). This decreases the chance of the message getting intercepted by an unintended user to protect the information being sent. Also links to recruit individuals have an expiration time (Bloom et al., 2019), encouraging potential recruits to join quickly. It is also a safety feature for the organisation in case the link is shared but not used; with an expiration time, it decreases the chance of the group getting caught by authorities. Bloom et al. (2019, p. 4) state there are three types of jihadist Telegram users: those who search for content, committed sympathisers and propagandists who are actively creating groups and official propagandists who create proxies. ISIS uses Telegram to participate in chatrooms which are used to recruit new members privately (Krona, 2020).

Two-way interactivity is present within radicalisation as audiences become active participants when they engage and respond to comments and chatrooms online (Aly et al., 2017). Multiple researchers highlight the advantages of these alt-tech platforms to extremists in the aim of radicalisation. Advantages to their success include free communication, the end-to-end encryption for heightened security and instant service of distributing material to a targeted audience (Gray & Head, 2009; Krona, 2020; Weimann, 2006). Engaging in networking is another way in which extremists utilise the Internet (Weimann, 2004). Weimann (2004) explains how the Internet provides violent extremist groups with instant connection around the world regardless of geographical location and a sense of readiness 24/7 due to push notifications. Global networks are created within large platforms due to their reach. Krona (2020) states the Internet gives the capability to extremist groups to operate as a more decentralised organisation as communication can be conducted over networks; this brings Weimann's (2004) remarks up to date. Social networks are not only used for the exchange of ideologies, but these anonymous platforms share instruction manuals on how to make bombs, poison and carry-out attacks (Weimann, 2004). Active users are encouraged to refer to these documents and act on them in the offline world. These 'how to' guides are too easily accessible within platforms such as Gab. With the addition of users viewing videos of 'successful' attacks due to homemade bombs, it could possibly be even more tempting for individuals to act on this to see if they can achieve what they see in the video, leading to becoming radicalised. Young men can become easily attracted to the visual imagery used in propaganda videos from ISIS, due to the real-life likeness of video games (Ali, 2015; Al-Rawi, 2018).

The propaganda videos consist of very high-quality cinematography used in propaganda videos, like a real-life video game. This kind of platform may be appealing to those who are searching for their identity, faith or sense of belonging (Baumeister & Leary, 1995). 'Successful' organised attacks are also portrayed through online magazines developed by terrorist groups. Online magazines are persuasive towards young men as they explain if women and children can provide support for their country, men should participate too. The ISIS magazine *Dabiq* convinces the reader to engage in acts of violent extremism and persuades individuals to travel to the Middle East. If they cannot commit to this, they are encouraged to perform lone-wolf attacks in their home country (Bertram, 2016). The al-Qaeda magazine *Inspire* is less informed and targeted towards less intellectual individuals which includes instruction manuals and drives readers to act. The fifth and sixth ways Weimann states terrorists use the Internet are known as mobilisation and coordination (2004). The Internet can be used to mobilise followers to become more involved in active roles to assist terrorist activity. Online communication also plays a role in this as it can enable extremist groups to coordinate members to undertake action. This action may be participating in demonstrations, rallies or engaging in violent extremist behaviour. Coordination of groups on smaller platforms can organise live streams to take place on mainstream platforms (Conway & Dillon, 2019). Wyman's six ways the Internet facilitates violent extremism are all intertwined and contribute together to the extremist behaviours they result in. In addition to Wyman's six factors, the use of social media to extremist's aim of radicalisation in the mainstream world such as Facebook and Twitter must be mentioned.

5　Social Media Platforms and Terrorism

Extremists use mainstream SM to increase their reach even further to the wider population (Schmid, 2013) because it is inexpensive, easily accessible and multi-media options are available such as video and image usage. The extremist organisations conduct the same acts on these platforms as they do with Gab—spread ideology, create fear within societies, motivate problems, recruit new members, display propaganda and provide an in-group/outgroup narrative. SMPs cannot do anything directly harmful to individuals or societies but can pose a threat on the outside world. The use of images being uploaded online is popular within extremist groups. Photographs are an easier tool for communication which gains quicker attention from the reader compared to words and creates social knowledge (Hariman & Lucaites, 2007). Viewers are also more likely to remember images and the meaning behind them better than text as there are no language barriers (Kovács, 2015) and gain a quicker and more positive emotional reaction (Goldberg & Gorn, 1987). Images are likely to be submerged into popular areas of SMPs to gain the maximum attraction. Extremist groups attach hashtags onto images to divert them into trending material online. For example, '#worldcup' was used for propaganda material to appear in the thread of trending tweets in the aim to recruit individuals into extremist groups (Milmo, 2014). The more views extremists gain on their online material, and their chance of radicalising individuals increase.

However, it could be deemed that SMPs could be ISIS' downfall. As profiles remain high in anonymity, those who claim they will act on extremist ideology may be hiding behind a keyboard and never act upon it. This explains the online disinhibition effect by Suler (2004), who states it is easier to speak with no disinhibition online, as communication is delivered differently in the offline world (Bjelopera, 2012). Therefore, ISIS may believe more people are committing acts compared to reality. Another downfall of social media for ISIS is the possibility that novice users may provide insight to counter terrorism due to their lack of knowledge within these platforms. Inquisitive new users may not be as reluctant as 'experts' within SMPs, which could lead them to accessing links taking them to an unknown source. There is a possibility that external links may be tracked by authorities which could be costly to extremist groups and, nevertheless, would be positive for the wider society.

Problems that arise with regulating SMPs include balancing censorship with freedom of speech. It must be remembered that individuals have the right to freedom of expression; however the rules on social media platforms such as Facebook's Community Standards (n.d.) state they must not include hate speech. There is also a lack of clear universal definition of terms such as extremism and terrorism across the globe. This can introduce a conflict of requirements within privacy and security when regulating social media (Home Office, 2011). Due to the increased reach available to extremist groups online, moderators struggle with protecting individuals within the wider community as they should not be seeing extremist content online. Terrorists and extremist groups are actively seeking new ways to take advantage of the Internet, and it seems moderators and governments are not as quick to access new tools. Thomas (2003, p. 114) also states how 'governments cannot control the Internet to the same degree they could control newspapers and TV'. There is also a problem within mainstream and alt-tech platforms. Take YouTube, for example; their algorithms of suggested videos could act as the pipeline to radicalisation due to videos being presented in front of them even when they do not go actively seeking for it. Individuals should report these videos when prompted with them, so they can be removed to decrease the chance of radicalisation.

6   The Internet of Things and Radicalisation

The IoT consists of devices connected to the Internet which consist of sensors, software and the ability to transfer data over a network without requiring human interaction (Gillis, 2020). Real-world examples of IoT devices include wearable smart watches, smart TVs, voice assistants such as Amazon Alexa and contactless payments (Anumala & Busetty, 2015). More recently in technological develop-ments, Artificial Intelligence of Things (AIoT) has become more widely used to achieve more efficient IoT operations (Christensen, 2019). Christensen (2019) suggests the IoT manages the devices connected to the Internet, while AI makes the device learn future tasks based on data and experience. AI can be advantageous in developing systems within the health setting, marketing and social media moderating. However, AI can be used for malicious purposes within the process of radicalisation towards extremism. Although it is expected that AI is extensively used to promote propaganda and spread extremist material online, the assumption is widely understudied and requires more empirical research. This is expected in the future due to the emergence wireless technology and the IoT becoming more pervasive (Hasim et al., 2016). Evidence suggesting the IoT does not encourage radicalisation derives from Schroeter (2020). The author suggests not enough data can be gathered on SM to create an algorithm which proves an individual has been radicalised online and committed terrorist acts. However, arguing against this, it has been shown that AI has taken place to distribute 'fake news'. This was conducted by creating realistic photographs and new accounts to distribute information which avoided detection from social media software which seeks false accounts (Villasenor, 2020). Although this was not related to terrorist material, it poses the threat that terrorist and extremist groups can do this in the future to avoid identity detection. AI is also used in the prevention of terrorist material being posted online via social media moderating. Keywords can be detected in posts which automatically remove the sensitive material. However, terrorist groups attempt to avoid the detection by inserting punctuation in-between words and by creating new accounts. However, AI can also falsely remove posts if the material is incorrectly detected as offensive; this remains a fault in the AI system which is expected to be corrected as technological advancements continue to develop.

7   The Internet of Things Moderating Online Terrorism

Countering online extremism has increased dramatically over the past years due to technology evolving and enabling moderators to remove content and protect the wider public. Decreasing the amount of extremist content online requires coordinated response across governments, private companies and independent regulators (Guelke, 2009). The Internet of Things plays a role aiming to stop extremist activity being posted online on social media platforms. AI and ML technologies are two ways that can be used to detect and remove activity without human interaction (Macdonald et al., 2019). AI and ML 'learn rules from data, adapt to changes, and improve performance with experience' (Blum, 2007, p. 1), whereas a content moderator is required to manually remove content. While technological and societal interventions are available and somewhat effective, these interferences alone are unlikely to eliminate terrorism entirely. Also, a study has found 4.8% of detected offensive tweets were misclassified; they contained offensive language which did not involve hateful words (Gaydhani et al., 2018). Although this was a small percentage of wrongful detection, the remaining 95.6% was accurately detected and therefore successfully removed. Since 2006, the British government announced its public strategy to counter international terrorism (Home Office, 2011). Their aim was to tackle terrorist use of the Internet as platforms are used to display many radical views, which can influence vulnerable individuals. One example of a team of human moderators who aim to remove online terrorist material is the Counter Terrorism Internet Referral Unit (CTIRU) in the UK. Established in 2010, they challenge the increase of extremist- and terrorist-related material posted online by removing or modifying content (Counter Terrorism Policing, 2018). Since 2010, they have successfully removed over 304,000 items of unlawful Internet content and continue to identify those liable for posting harmful material. The unit actively scans the Internet for extremist content as well as researching into reported websites by the public. Automatic content removal using AI and ML algorithms is a popular intervention for reducing terrorist-related material on social media as it can be removed quickly and reduces the number of users viewing extremist material. However, sometimes it is not removed quickly enough. If content is removed successfully and efficiently, this safeguards human moderators from viewing harmful material as it could be psychologically damaging. The difficulty in automatically removing posts from word detection is the range of languages the messages can be displayed in. Social media companies would have to recruit bilingual speakers to detect hate speech and extremist posts in different languages. Also, experienced users can avoid word detection not using specific terms that are usually recognised and removed. This makes it difficult for automatic detection to be efficient, and therefore content is accessible for longer allowing more people to view the material. Users also post multi-media such as videos, images and memes to avoid getting detected. Also, if material is posted on the dark web, it can be impossible to remove due to the decentralised server. There are other ways online extremism can be countered such as de-platforming, societal and individual interventions.

8   Conclusion

It can be stated that the Internet can play a sole role in radicalising individuals into violent extremism as offline external factors are not always required; however, they can lead to social events in the offline world such as protest marches through the communication of online advertisement. The role of the Internet most definitely affects radicalisation in individuals into violent extremism described through the explanation from Weimann (2004). The continuation of countering violent extremist radicalisation is required with the integrated help of authorities such as the CTIRU and social media platform regulators with the heuristic aim to decrease violent extremist behaviours and terrorist acts. Although it will not eliminate terrorism entirely, it will be a huge contribution. The idea of radicalisation is very complex, and its diversity ranges from case to case in terms of how individuals are radicalised. Critiques assume radicalisation is wholly assumption and intuition based as it is not scientific empirically based research. This area of violent extremism requires more research for it to be deemed empirical evidence which could consequently reduce the risks of terrorist acts occurring in the future.

References

Ali, M. (2015). *ISIS and propaganda: How ISIS exploits women* (pp. 10–11). Reuters Institute for the Study of Journalism.

Al-Rawi, A. (2018). Video games, terrorism, and ISIS's Jihad 3.0. *Terrorism and Political Violence, 30*(4), 740–760.

Aly, A., Macdonald, S., Jarvis, L., & Chen, T. M. (2017). Introduction to the special issue: Terrorist online propaganda and radicalization. *Studies in Conflict & Terrorism, 40*(1), 1–9.

Amedie, J. (2015). The impact of social media on society. *Advanced Writing: Pop Culture Intersections, 2.*

Anglano, C., Canonico, M., & Guazzone, M. (2017). Forensic analysis of telegram messenger on android smartphones. *Digital Investigation, 23,* 31–49.

Anumala, H., & Busetty, S. M. (2015). Distributed device health platform using internet of things devices. In *2015 IEEE International Conference on Data Science and Data Intensive Systems* (pp. 525–531).

Archetti, C. (2013). Terrorism, communication, and the media. In *Understanding terrorism in the age of global media* (pp. 32–59).

Baruah, T. D. (2012). Effectiveness of Social Media as a tool of communication and its potential for technology enabled connections: A micro-level study. *International Journal of Scientific and Research Publications, 2*(5), 1–10.

Baugut, P., & Neumann, K. (2020). Online news media and propaganda influence on radicalized individuals: Findings from interviews with Islamist prisoners and former Islamists. *New Media & Society, 22*(8), 1437–1461.

Baumeister, R. F., & Leary, M. R. (1995). The need to belong: Desire for interpersonal attachments as a fundamental human motivation. *Psychological Bulletin, 117*(3), 497.

Benigni, M. C., Joseph, K., & Carley, K. M. (2017). Online extremism and the communities that sustain it: Detecting the ISIS supporting community on Twitter. *PLoS One, 12*(12), e0181405.

Bertram, L. (2015). How could a terrorist be de-radicalised? *Journal for Deradicalization, 5,* 120–149.

Bertram, L. (2016). Terrorism, the Internet and the social media advantage: Exploring how terrorist organizations exploit aspects of the internet, social media and how these same platforms could be used to counter-violent extremism. *Journal for Deradicalization, 7,* 225–252.

Bjelopera, J. P. (2012). *The domestic terrorist threat: Background and issues for Congress.* CRS report for Congress.

Bloom, M., Tiflati, H., & Horgan, J. (2019). Navigating ISIS's preferred platform: Telegram1. *Terrorism and Political Violence, 31*(6), 1242–1254.

Blum, A. (2007). *Machine learning theory* (p. 26). Carnegie Melon University, School of Computer Science.

Borum, R. (2014). Psychological vulnerabilities and propensities for involvement in violent extremism. *Behavioral Sciences & the Law, 32*(3), 286–305.

Butler, A. S., Panzer, A. M., & Goldfrank, L. R. (2003). Understanding the psychological consequences of traumatic events, disasters, and terrorism. In *Preparing for the psychological consequences of terrorism: A public health strategy.* The National Academies Press.

Christensen, G. (2019, February). *Artificial Intelligence of Things (AIoT).* IoT Agenda. Retrieved from https://internetofthingsagenda.techtarget.com/definition/Artificial-Intelligence-of-Things-AIoT

Christmann, K. (2012). *Preventing religious radicalisation and violent extremism: A systematic review of the research evidence.* Youth Justice Board.

Cohen-Almagor, R. (2005). Media coverage of terror: Troubling episodes and suggested guidelines. *Canadian Journal of Communication, 30*(3), 383–409.

Conway, M., & Dillon, J. (2019). Future trends: Live-streaming terrorist attacks? *VOX-Pol.*

Counter Terrorism Policing. (2018, April 6). *Specialist unit tackles online extremism.* Retrieved from https://www.counterterrorism.police.uk/specialist-unit-tackles-online-extremism/

Crenshaw, M. (1981). The causes of terrorism. *Comparative Politics, 13*(4), 379–399.

Crosby, F. (1976). A model of egoistical relative deprivation. *Psychological Review, 83*(2), 85.

D'Incau, F., & Soesanto, S. (2017). Countering online radicalisation [Blog]. Retrieved from https://ecfr.eu/article/commentary_countering_digital_radicalisation_7216/

D'Souza, S. M. (2015). Online radicalisation and the specter of extremist violence in India. *Mantraya Brief, 1*.

Dekel, R., & Nuttman-Shwartz, O. (2009). Posttraumatic stress and growth: The contribution of cognitive appraisal and sense of belonging to the country. *Health & Social Work, 34*(2), 87–96.

Diener, E. (1979). Deindividuation, self-awareness, and disinhibition. *Journal of Personality and Social Psychology, 37*(7), 1160.

Donovan, J., Lewis, B., & Friedberg, B. (2018). Parallel ports. Sociotechnical change from the Alt-Right to Alt-Tech. In *Post-digital cultures of the far right* (pp. 49–66). Transcript-Verlag.

Doosje, B., Moghaddam, F. M., Kruglanski, A. W., De Wolf, A., Mann, L., & Feddes, A. R. (2016). Terrorism, radicalization and de-radicalization. *Current Opinion in Psychology, 11*, 79–84.

Doosje, B., van den Bos, K., Loseman, A., Feddes, A. R., & Mann, L. (2012). "My in-group is superior!": Susceptibility for radical right-wing attitudes and behaviors in Dutch youth. *Negotiation and Conflict Management Research, 5*(3), 253–268.

Douglas, K. (2010). Deindividuation. *Encyclopaedia Britannica*. Retrieved from https://www.britannica.com/topic/deindividuation#ref310686

Facebook Community Standards. (n.d.). Objectional content. Hate speech. Community Standards. Retrieved from facebook.com

Ganor, B. (2002). Defining terrorism: Is one man's terrorist another man's freedom fighter? *Police Practice and Research, 3*(4), 287–304.

Gaydhani, A., Doma, V., Kendre, S., & Bhagwat, L. (2018). Detecting hate speech and offensive language on twitter using machine learning: An n-gram and TFIDF based approach. *arXiv preprint arXiv:1809.08651*.

Gill, P. (2007). A multi-dimensional approach to suicide bombing. *International Journal of Conflict and Violence (IJCV), 1*(2), 142–159.

Gill, P., Horgan, J., & Deckert, P. (2014). Bombing alone: Tracing the motivations and antecedent behaviors of lone-actor terrorists. *Journal of Forensic Sciences, 59*(2), 425–435.

Gillis, A. (2020). Internet of things. IoT agenda. What is IoT (Internet of Things) and how does it work? Retrieved from techtarget.com

Goldberg, M. E., & Gorn, G. J. (1987). Happy and sad TV programs: How they affect reactions to commercials. *Journal of Consumer Research, 14*(3), 387–403.

Goodwin, M. J. (2011). *New British fascism: Rise of the British national party*. Routledge.

Gray, D. H., & Head, A. (2009). The importance of the Internet to the post-modern terrorist and its role as a form of safe haven. *European Journal of Scientific Research, 25*(3), 396–404.

Grusin, R. (2010). *Premediation: Affect and mediality after 9/11*. Springer.

Guelke, A. (2009). *The new age of terrorism and the international political system*. IB Tauris.

Hardy, K. (2018). Comparing theories of radicalisation with countering violent extremism policy. *Journal for Deradicalization, 15*, 76–110.

Hariman, R., & Lucaites, J. L. (2007). *No caption needed: Iconic photographs, public culture, and liberal democracy*. University of Chicago Press.

Hasim, N. N. M., Mohamed, H., & Ibrahim, J. (2016). The effect and challenges of online radicalization on modern day society. *International Journal of Information and Communication Technology, 6*(12).

Hogg, M. A. (2020). Uncertain self in a changing world: A foundation for radicalisation, populism, and autocratic leadership. *European Review of Social Psychology, 32*, 235–268.

Hogg, M. A., Kruglanski, A., & Van den Bos, K. (2013). Uncertainty and the roots of extremism. *Journal of Social Issues, 69*(3), 407–418.

Home Office. (2011). *The prevent strategy* (pp. 107–108). Crown Publishing Group.

Hudson, R. A. (1999). *The sociology and psychology of terrorism: Who becomes a terrorist and why?* Library of Congress Washington Dc Federal Research Division.

Hussin, S. (2018). *Singapore's approach to countering violent extremism. Combatting violent extremism and terrorism in Asia and Europe* (p. 171). Konrad Adenauer Stiftung.

Jones, S. (2009). Radicalisation in Denmark. *Renewal, 17*(1), 22–28.

Jones, S. (2018). Radicalisation in the Philippines: The Cotabato Cell of the "East Asia Wilayah". *Terrorism and Political Violence, 30*(6), 933–943.

Jowett, G. S., & O'Donnell, V. (2018). *Propaganda & persuasion*. Sage Publications.

Koehler, D. (2014). The radical online: Individual radicalization processes and the role of the Internet. *Journal for Deradicalization, 1*, 116–134.

Kovács, A. (2015). The 'new jihadists' and the visual turn from al-Qa'ida to ISIL/ISIS/Da'ish. *Bitzpol Affairs, 2*(3), 47–69.

Krona, M. (2020). Collaborative media practices and interconnected digital strategies of Islamic state (IS) and pro-IS supporter networks on telegram. *International Journal of Communication, 14*, 1888–1910.

Kundnani, A. (2012). Radicalisation: The journey of a concept. *Race & Class, 54*(2), 3–25.

Macdonald, S., Correia, S. G., & Watkin, A. L. (2019). Regulating terrorist content on social media: Automation and the rule of law. *International Journal of Law in Context, 15*(2), 183–197.

McCauley, C., & Moskalenko, S. (2008). Mechanisms of political radicalization: Pathways toward terrorism. *Terrorism and Political Violence, 20*(3), 415–433.

Milmo, C. (2014). Iraq crisis exclusive: Isis jihadists using World Cup and Premier League hashtags to promote extremist propaganda on Twitter. *The Independent*.

Moghaddam, F. M. (2008). *How globalization spurs terrorism: The lopsided benefits of "one world" and why that fuels violence*. Praeger Security International.

Moghaddam, F. M., Heckenlaible, V., Blackman, M., Fasano, S., & Dufour, D. J. (2016). Globalization and terrorism. In *The social psychology of good and evil* (p. 415).

Neo, L. S., Dillon, L., & Khader, M. (2017). Identifying individuals at risk of being radicalised via the internet. *Security Journal, 30*(4), 1112–1133.

Patrikarakos, D. (2018). Web 2.0: The new battleground. *Armed Conflict Survey, 4*(1), 51–64.

Radicalisation Awareness Network. (2017). Working with families and safeguarding children from radicalisation. Retrieved from https://bit.ly/2UEP29a

Ramakrishna, K. (2011). Self-radicalisation and the Awlaki connection. In *Strategic currents* (pp. 140–142). ISEAS Publishing.

Sageman, M. (2004). *Understanding terror networks*. University of Pennsylvania Press.

Sageman, M. (2008). A strategy for fighting international Islamist terrorists. *The Annals of the American Academy of Political and Social Science, 618*(1), 223–231.

Salahuddin, M., & Alam, K. (2015). Internet usage, electricity consumption and economic growth in Australia: A time series evidence. *Telematics and Informatics, 32*(4), 862–878.

Schmid, A. P. (2013). Radicalisation, de-radicalisation, counter-radicalisation: A conceptual discussion and literature review. *ICCT Research Paper, 97*(1), 22.

Schroeter, M. (2020). Global network on extremism & technology. Artificial intelligence and countering violent extremism: A primer. Retrieved form https://gnet-research.org/wp-content/uploads/2020/10/GNET-Report-Artificial-Intelligence-and-Countering-Violent-Extremism-A-Primer_V2.pdf

Silber, M. D., Bhatt, A., & Analysts, S. I. (2007). *Radicalization in the West: The homegrown threat* (pp. 1–90). Police Department.

Smith, B. L. (2020) Propaganda. Encyclopaedia Britannica. Propaganda | definition, history, techniques, examples, & facts | Britannica.

Spears, R., Lea, M., & Lee, S. (1990). De-individuation and group polarization in computer-mediated communication. *British Journal of Social Psychology, 29*(2), 121–134.

Speckhard, A., & Ellenberg, M. (2020). Is internet recruitment enough to seduce a vulnerable individual into terrorism. *Homeland Security Today*.

Statista. (2020). Global digital population as of October 2020 [Graph]. Retrieved from https://www.statista.com/statistics/617136/digital-population-worldwide/

Stewart, B. B., & Thompson, J. W. (2002). U.S. Patent No. 6,414,635. U.S. Patent and Trademark Office, Washington, DC. 1498394371873592871-06414635 Retrieved from storage.googleapis.com

Striegher, J. L. (2015). *Violent-extremism: An examination of a definitional dilemma*. Edith Cowan University.

Suler, J. (2004). The online disinhibition effect. *Cyberpsychology & Behavior, 7*(3), 321–326.

Tajfel, H., & Turner, J. (1986). The social identity theory of intergroup behavior. In J. T. Jost & J. Sidanius (Eds.), *Political psychology: Key readings* (pp. 276–293). Psychology Press.

Thomas, T. L. (2003). *Al Qaeda and the Internet: The danger of 'Cyberplanning'*. Foreign Military Studies Office (ARMY).

Torok, R. (2013). Developing an explanatory model for the process of online radicalisation and terrorism. *Security Informatics, 2*(1), 6.

Trip, S., Bora, C. H., Marian, M., Halmajan, A., & Drugas, M. I. (2019). Psychological mechanisms involved in radicalization and extremism. A rational emotive behavioral conceptualization. *Frontiers in Psychology, 10*, 437.

Tzezana, R. (2016). Scenarios for crime and terrorist attacks using the internet of things. *European Journal of Futures Research, 4*(1), 18.

United Nations Office on Drugs and Crime. (2012). *Use of the internet for terrorist purposes*. United Nations Office on Drugs and Crime.

Van den Bos, K. (2018). *Why people radicalize: How unfairness judgments are used to fuel radical beliefs, extremist behaviors, and terrorism*. Oxford University Press.

Villasenor, J. (2020). *How to deal with AI-enabled disinformation*. Brookers.

Vincent, J. (2019). Facebook bans UK's biggest far-right organizations, including EDL, BNP, and Britain First [Blog]. Retrieved from https://www.theverge.com/2019/4/18/18484623/facebook-bans-uk-far-right-groups-leaders-edl-bnp-britain-first

Weimann, G. (2004). *www.terror.net: How modern terrorism uses the Internet* (Vol. 31). United States Institute of Peace.

Weimann, G. (2006). Virtual disputes: The use of the Internet for terrorist debates. *Studies in Conflict & Terrorism, 29*(7), 623–639.

Weimann, G. (2014). *New terrorism and new media* (Vol. 2). Commons Lab of the Woodrow Wilson International Center for Scholars.

Weimann, G. (2016). Going dark: Terrorism on the dark web. *Studies in Conflict & Terrorism, 39*(3), 195–206.

Wibisono, S., Louis, W. R., & Jetten, J. (2019). A multi-dimensional analysis of religious extremism. *Frontiers in Psychology, 10*, 2560.

Wiktorowicz, Q. (Ed.). (2004). *Islamic activism: A social movement theory approach*. Indiana University Press.

The Internet of Things and Terrorism: A Cause for Concern

Joseph Rees

1 Introduction

Over recent years, there have been rapid advances in information and communication technology. The Internet of Things (IoT), an instance of such technologies, has brought numerous benefits to societies, revolutionising the lifestyles of many individuals living in these societies. Whilst advances in the IoT undoubtedly offer numerous benefits, they simultaneously present a wide range of new security threats that can have devastating impacts on societies (Montasari & Hill, 2019). Although research in the field of counter terrorism frequently discusses the use of technology by terrorists, the focus predominantly centres on platforms and the utilisation of the Internet, in areas of anonymity, communication and networking, dissemination of propaganda, recruitment and financing (English, 2010). These concerns are heavily documented, as is the growing concern with the misuse of emerging technologies by groups including terrorist organisations (Lubrano, 2021). In the field of emerging technologies, it has been noted that Internet of Things (IoT) devices with obvious security connotations such as smart alarm systems and smart locks; however, the increasingly wide range of these devices also includes items which, in the first instance, may be overlooked as far as relevance to terrorism and counterterrorism. Such items may include voice controllers, smart doorbells, smart smoke alarms, smart watches, smart fridges, air quality monitors, home Wi-Fi systems and smart bicycles. It should be noted that these devices also contain a vast array of operating systems and systems architecture (Stoyanova et al., 2020). It is not within the remit of this chapter to consider these types of devices on an item-by-item basis. Rather, the purpose of this chapter is to highlight and discuss the introduction of Internet

J. Rees (✉)
Hillary Rodham Clinton School of Law, Swansea University, Swansea, UK
e-mail: 914770@Swansea.ac.uk; http://www.swansea.ac.uk

© The Author(s), under exclusive license to Springer Nature Switzerland AG 2022
R. Montasari et al. (eds.), *Privacy, Security And Forensics in The Internet of Things (IoT)*, https://doi.org/10.1007/978-3-030-91218-5_10

of Things (IoT) devices into everyday life (Khan & Salah, 2018) and the potential implications of this trend for terrorism and hence counterterrorism.

The remainder of the chapter is structured as follows. First the emergence of new terrorism in the post-9/11 era is highlighted. Aspects of this new terrorism are then discussed with reference to IoT. The chapter concludes by identifying the challenge of IoT for those involved in counterterrorism activities.

2   Trends in Terrorism

By its very nature, terrorism is an ever-evolving threat, and there is an established body of literature which has explored patterns and trends relating to both terrorism and counterterrorism (Hoffman, 2010). Notably, within this body of research, the events of 9/11 are consistently identified as a pivotal point between what has been termed 'old' and 'new' terrorism (Jensen, 2009). That is, a significant pattern in terrorism is believed to have emerged post 9/11. This attack, still described as the most lethal terrorist attack in history (FBI, 2021), is one which the 9/11 commission has highlighted as the dawn of the new era of terrorism (US Gov, 2004). Although some continuity exists across the ideologies, methods and organisational structures of both the 'old' and the 'new' (Field, 2009), three main aspects of new terrorism have been identified, that is its primary use of apocalyptic religious ideologies; its bloodier, unrestrained methods; and finally the decentralisation of new terrorist groups (Crenshaw, 2003). These three aspects of new terrorism are discussed below with reference to IoT considerations.

3   Apocalyptic Aspects of New Terrorism

In line with Rapoport's (2001) fourth wave of terrorism, religion is said to be central to the ideologies of new terrorism. The fear and explanations of 'religious' attacks often lie in the so-called 'apocalyptic' nature of these ideologies. Perhaps the move of terrorism away from political agendas (Jenkins, 2006) represents a return to a time when religion was seen by some as the 'only accepted justification for terror' (Rapoport, 1983, p. 659) especially in Western societies in which political violence is simply not tolerated. Furthermore, events such as the 1993 attack on the World Trade Center could be seen as a precursor to 9/11 and a sign of slow changes that were already happening in the early 1990s. During ISIS' online campaigns, the utilisation of online platforms, in particular Twitter, for propaganda purposes highlighted the ability for terrorist groups to leverage existing technology in order to further their cause. What was also seen during the Twitter campaign was the creation of the mobile app 'The Dawn of Glad Tidings' in order to circumnavigate Twitter's spam detection algorithms. The potential of IoT devices being used in such a way is certainly a possibility.

4   Bloodier Terrorism

A further observation that can be made about the so-called new era of terrorism relates to its 'apocalyptic' nature as discussed above. That is, it has been argued that terrorism become bloodier due to absolutism and lack of compromise. This argument gives rise to a series of key questions. For example, has there been a vast transformation from the traditional 'propaganda of the deed' (Fleming, 1980) to an era in which the 'means have become an end in themselves' (Crenshaw, 2003)? Or, again, is the insinuation that the religious absolutism of 'new terrorists' is something new ignoring the reflective trends of earlier terrorist groups (Copeland, 2001)?

Scholars have sought to argue that terrorism has become bloodier over time (Jenkins, 2006) with earlier 'milder' terrorism possibly attributable to the 'sensible' temperament of the old terrorists (Crenshaw, 2003, p. 1). In a thought-provoking contribution, Laqueur (1998) has argued that older forms of terrorism often followed a strict list of implicit rules. In contrast, more contemporary indiscriminate and ruthless tactics such as suicide bombings are associated with new terrorism (Crenshaw, 2003). Yet the use of tactics such as suicide bombing is not new. For example, suicide bombing was carried out prior to 9/11 in the 1980s in Lebanon (Horowitz, 2015). Similarly, early adopters such as Liberation Tigers of Tamil Eelam (LTTE) frequently used this bloody tactic. Unlike militaries who require technology in order to establish superiority, terrorist organisations will utilise any available technology, however high or low tech, if it helps further their cause. Thus, even attrition tactics utilising low-end technology can be considered useful for terrorist organisations. For example, the Casio F-91W digital watch in the fabrication of bombs by al-Qaeda was used so heavily it became known as the 'sign of al-Qaeda'. The question of whether terrorists will look to use technology is arguably a rhetorical question; based on the abundance and incorporation of IoT devices into society, it is likely that terrorists seek to make use of these devices for activities such as surveillance, weapon fabrication and communication. Rather, more vexatious questions surround the potential utility of new IoT devices for those engaged in terrorist activity.

In summary, at a pragmatic level, current research offers little evidence about, or even discussion of, the acquisition and use of IoT devices by terrorists. At a more theoretical level, this observation also raises questions about whether the older forms of terrorism simply lacked modern technology to enable terrorists to achieve more devastating results. Arguably, had modern IoT technology, including sophisticated communication networks, been developed earlier, bloody terrorism would have been far more widespread prior to 9/11.

5   Decentralisation

A third consideration associated with new terrorism is said to be its unpredictability and lack of accountability due to its decentralised nature (Crenshaw, 2003). For example, scholars have highlighted al-Qaeda's ability to operate across national borders (Boeke, 2016), a process facilitated by the unrivalled global reach of the Internet. It is recognised that modern contemporary groups were utilising transnational terrorism (Sandler & Enders, 2004) as early as 1968. Jensen (2008) even draws similarities of the global scope of al-Qaeda to the anarchists of the late nineteenth century. More specifically, the issue of decentralisation and new terrorism becomes more focused in relation to the funding of terrorist groups. That is, it has been argued that new terrorism is now less reliant on state sponsors due to the decentralisation of terrorist groups over more recent years (Morgan, 2004).

The LTTE in Sri Lanka is one example of a group that received funds from global sources. One of the consequences of 9/11 was that state sponsorship of terrorism attracted increasing levels of attention and accountability on the global stage. What is not clear is whether the decentralisation of terrorist organisations has been mainly caused by the drying up of funds from state sponsors since 9/11 or whether this decentralisation represents a strategy designed by terrorists to make their groups more resilient to, for example, the loss of members. Nevertheless, although decentralisation signifies less formal hierarchical structures within terrorist organisations, it also involves increasingly complex intra- and intercommunication networks both within and between terrorist organisations. As previously noted, this communication has relied on both established and ever-evolving technological means ranging from handheld radios to encrypted messaging applications. Terrorists will inevitably draw on both existing IoT devices and IoT devices currently in development, such as new devices using existing protocols such as 'Voice over Internet Protocol' to enhance their modes of communication and to evade counterterrorism efforts.

6 Discussion

Since 9/11, there has been a significant increase in the role of the Internet in individual radicalisation (Koehler, 2014). Unfortunately, terrorists are known to 'innovate; exploit new technology; learn from one another; imitate successful tactics; produce manuals of instruction based on experience; debate tactics, targets, and limits on violence; and justify their actions with doctrines and theories' (Jenkins, 2006, p. 117). In essence, just as terrorists have, in the past, utilised technologies such as the telephone in order to facilitate conventional terrorism, it would indeed 'be strange' if they did not utilise the Internet (Benson, 2014) and, therefore, IoT devices. It is relevant to highlight that, at a broad level, researchers have clearly identified the potential and actual use of the Internet by terrorists. For example, Weimann (2006, p. 51) has stated that 'the Internet appeals to terrorists for the same reasons it attracts everyone else: it is inexpensive, easily accessible, has little or no regulation, is interactive, allows for multimedia content, and the potential audience is huge. And it's anonymous'.

The emergence of IoT devices, however, calls for a re-examination of how terrorists can engage with the Internet. These devices may fall outside of mainstream considerations, yet they have the potential to be used nefariously by those with ill intent. More recently, academics have highlighted the security concerns related to IoT devices from the governmental policy level (Crawford & Sherman, 2018) to general surveys of IoT security in regard to systems architecture (Mendez Mena et al., 2018). Academics have also gone further by undertaking risk assessments and providing potential frameworks for the management of IoT security (Dhar & Bose, 2021). However, the rapidity of the development and complexity of these devices, coupled with their increasingly ubiquitous nature, now represent a major challenge to those involved in counterterrorism.

7   Conclusion

As IoT devices become increasingly incorporated into society, then it inevitable that terrorists will utilise this technology. That is, as IoT expands so will its utilisation by terrorists. Back in 2006, Weimann (2006, p. 78) stated that terrorist behaviour is: '. . . not only proficient but also imaginative and innovative'. More recently, Torres-Soriano (2021) concluded that technological barriers to entry were the most significant factors when considering terrorist activists' usage of online spaces. The role of the Internet in terrorism (Conway, 2017) has been established in recent years; yet the promising future of IoT for society also represents a threat from a terrorism perspective. This threat will be difficult to address by counterterrorism agencies given with the wide range of IoT hardware and software referred to in the introduction of this chapter. Addressing this threat will require heightened levels of awareness by law enforcement agencies and perhaps also the developers of these devices.

References

Benson, D. C. (2014). Why the Internet is not increasing terrorism. *Security Studies, 23*(2), 293–328.

Boeke, S. (2016). Al Qaeda in the Islamic Maghreb: Terrorism, insurgency, or organized crime? *Small Wars & Insurgencies, 27*(5), 914–936.

Conway, M. (2017). Determining the role of the internet in violent extremism and terrorism: Six suggestions for progressing research. *Studies in Conflict & Terrorism, 40*(1), 77–98.

Copeland, T. (2001). Is the "new terrorism" really new?: An analysis of the new paradigm for terrorism. *Journal of Conflict Studies, 21*(2), 7–27.

Crawford, D., & Sherman, J. (2018). Gaps in United States federal government IoT security and privacy policies. *Journal of Cyber Policy, 3*(2), 187–200.

Crenshaw, M. (2003). Is today's "new" terrorism qualitatively different from pre-September 11 "old" terrorism? *Palestine-Israel Journal, 10*(1).

Dhar, S., & Bose, I. (2021). Securing IoT devices using zero trust and blockchain. *Journal of Organizational Computing and Electronic Commerce, 31*(1), 18–34.

English, R. (2010). *Terrorism: How to respond.* Oxford University Press.

FBI. (2021). 9/11 Investigation. Retrieved July 17, 2021, from https://www.fbi.gov/history/famous-cases/911-investigation

Field, A. (2009). The 'new terrorism': Revolution or evolution? *Political Studies Review, 7*(2), 195–207.

Fleming, M. (1980). Propaganda by the deed: Terrorism and anarchist theory in late nineteenth-century Europe. *Studies in Conflict & Terrorism, 4*(1–4), 1–23.

Hoffman, B. (2010). The evolving nature of terrorism—Nine years after the 9/11 attacks. Retrieved July 15, 2021, from https://www.govinfo.gov/content/pkg/CHRG-111hhrg66029/html/CHRG-111hhrg66029.htm

Horowitz, M. C. (2015). The rise and spread of suicide bombing. *Annual Review of Political Science, 18*, 69–84.

Jenkins, B. M. (2006). The new age of terrorism. In *The McGraw-Hill homeland security handbook* (pp. 117–130).

Jensen, R. B. (2008). Nineteenth century anarchist terrorism: How comparable to the terrorism of Al-Qaeda? *Terrorism and Political Violence, 20*(4), 589–596.

Jensen, R. B. (2009). The international campaign against anarchist terrorism, 1880–1930s. *Terrorism and Political Violence, 21*(1), 89–109.

Khan, M. A., & Salah, K. (2018). IoT security: Review, blockchain solutions, and open challenges. *Future Generation Computer Systems, 82*, 395–411.

Koehler, D. (2014). The radical online: Individual radicalization processes and the role of the Internet. *Journal for Deradicalization, 1*, 116–134.

Laqueur, W. (1998). The new face of terrorism. *Washington Quarterly, 21*(4), 167–178.

Lubrano, M. (2021). Stop the machines: How emerging technologies are fomenting the war on civilization. *Terrorism and Political Violence*, 1–17.

Mendez Mena, D., Papapanagiotou, I., & Yang, B. (2018). Internet of things: Survey on security. *Information Security Journal: A Global Perspective, 27*(3), 162–182.

Montasari, R., & Hill, R. (2019). Next-generation digital forensics: Challenges and future paradigms. In *2019 IEEE 12th International Conference on Global Security, Safety and Sustainability (ICGS3)* (pp. 205–212). IEEE.

Morgan, M. J. (2004). The origins of the new terrorism. Military Intelligence BN (125th) Schofield Barracks HI. Retrieved July 17, 2021, from https://apps.dtic.mil/sti/citations/ADA597084

Rapoport, D. C. (1983). Fear and trembling: Terrorism in three religious traditions. *American Political Science Review, 78*(3), 658–677.

Rapoport, D. C. (2001). The fourth wave: September 11 in the history of terrorism. *Current History, 100*(650), 419–424.

Sandler, T., & Enders, W. (2004). An economic perspective on transnational terrorism. *European Journal of Political Economy, 20*, 301–316.

Stoyanova, M., Nikoloudakis, Y., Panagiotakis, S., Pallis, E., & Markakis, E. K. (2020). A survey on the internet of things (IoT) forensics: Challenges, approaches, and open issues. *IEEE Communications Surveys & Tutorials, 22*(2), 1191–1221.

Torres-Soriano, M. R. (2021). Barriers to entry to Jihadist activism on the Internet. *Studies in Conflict & Terrorism*, 1–18.

United States Government. (2004). The 9/11 Commission Report. Retrieved July 15, 2021, from https://www.9-11commission.gov/report/911Report.pdf

Weimann, G. (2006). *Terror on the Internet: The new arena, the new challenges*. US Institute of Peace Press.

Correction to: Privacy and Security Challenges and Opportunities for IoT Technologies During and Beyond COVID-19

V. Bentotahewa, M. Yousif, C. Hewage, L. Nawaf, and J. Williams

Correction to:
Chapter 3 in: R. Montasari et al. (eds.),
Privacy, Security And Forensics in The Internet of Things (IoT),
https://doi.org/10.1007/978-3-030-91218-5_3

The book was inadvertently published with an incorrect name information for one of the Chapter author as "M. Yousef", whereas it should be "M. Yousif" in the front matter and Chapter 3.

The updated online version of this chapter can be found at
https://doi.org/10.1007/978-3-030-91218-5_3

Index

Printed in the United States
by Baker & Taylor Publisher Services